To Professor Ian Hodder

Luca Cavalli-Sforza

2-9-00

THE
NEOLITHIC TRANSITION
AND THE GENETICS
OF POPULATIONS IN
EUROPE

THE NEOLITHIC TRANSITION AND THE GENETICS OF POPULATIONS IN EUROPE

ALBERT J. AMMERMAN
AND
L. L. CAVALLI-SFORZA

PRINCETON UNIVERSITY PRESS
PRINCETON, NEW JERSEY

Copyright © 1984 by Princeton University Press
Published by Princeton University Press,
41 William Street, Princeton, New Jersey 08540
In the United Kingdom:
Princeton University Press, Guildford, Surrey

ALL RIGHTS RESERVED

Library of Congress Cataloging in Publication Data
will be found on the last printed page of this book
ISBN 0-691-08357-6

This book has been composed in Linotron Baskerville
Clothbound editions of Princeton University Press books
are printed on acid-free paper, and binding materials
are chosen for strength and durability

Printed in the United States of America by
Princeton University Press,
Princeton, New Jersey

Dedicated to the memory of Adriano Buzzati-Traverso,
to whose inspiring teaching this research owes its source.

CONTENTS

	Figures	ix
	Tables	xi
	Preface	xiii
1	Introduction	3
2	The Origins of Agriculture	
	2.1 Introduction	9
	2.2 Multiple Centers of Origin	13
	2.3 Plant Domestication	16
	2.4 Animal Domestication	21
	2.5 Explaining the Origins of Agriculture	24
	2.6 The Growth of Food Production	30
3	The Neolithic Transition in Europe	
	3.1 European Prehistory on Its Own Terms	34
	3.2 The Mesolithic Background	35
	3.3 Early Farming Cultures in Europe	39
	3.4 The Neolithic Transition	45
4	Measuring the Rate of Spread	
	4.1 The Initial Analysis	51
	4.2 Further Analysis	58
	4.3 Interpreting the Rate of Spread	60
5	The Wave of Advance Model	
	5.1 The Neolithic Rise in Population	63
	5.2 The Model	67
	5.3 Logistic Growth	71
	5.4 Migratory Activity	76
	5.5 An Initial Test	80
	5.6 Genetic Implications of the Model	82
6	The Analysis of Genes	
	6.1 Expectations for the Geographic Distribution of Genes	85
	6.2 The Rh Gene	86
	6.3 Mechanisms of Evolutionary Change	87
	6.4 Evidence from Other Genes	93

	6.5	Toward a Synthetic View	99
	6.6	Principal Components Analysis and Synthetic Maps	102
	6.7	The Principal Components Maps	105
7	Simulation Studies		
	7.1	Toward the Study of Process	109
	7.2	Simulation Study of Settlement Patterns	113
	7.3	Modeling Population Interactions	116
	7.4	Simulation Study of Genetic Patterns	119
	7.5	Evaluating Principal Components Analysis as a Method	130
8	Conclusions		133
	Appendix		141
	Site List 4.1 The neolithic sites with C-14 dates used to make the isochron map in figure 4.5 and with geographic coordinates and laboratory numbers	142
	Site List 4.2 The mesolithic sites with C-14 dates used to make the isochron map in figure 4.6 and with geographic coordinates and laboratory numbers	146
	Notes		149
	Bibliography		161
	Index		171

FIGURES

1.1	Distribution maps of hunters and gatherers	5
2.1	Centers of agricultural origins in the world	13
2.2	Diagram of a cereal spike	18
2.3	Distribution map of wild emmer in the Near East	21
2.4	Sites in the Near East	25
3.1	The production of geometric microliths	37
3.2	Selected mesolithic sites in Europe	38
3.3	Selected neolithic sites in Europe	40
3.4	Examples of early neolithic pottery in Europe	42
3.5	Distribution map of Bandkeramik sites near Aldenhoven	44
4.1	Map of early farming sites published by Clark in 1965	51
4.2	Regression analysis of the spread of early farming	53
4.3	Expected patterns of diffusion under four hypotheses	54
4.4	Radiocarbon dates and calendric dates	56
4.5	Isochron map of the spread of early farming in Europe	59
4.6	Isochron map of the "latest" mesolithic occupation	60
5.1	Life expectancy at 15 years of age	65
5.2	Fisher's model of a population wave of advance	69
5.3	The spread of the muskrat in central Europe	70
5.4	Two models of population growth	72
5.5	Local population densities and local net growth rates	74
5.6	Heuristic analysis of population growth at Aldenhoven	75
5.7	The treatment of migratory activity in Fisher's model	76
5.8	Distances between the birthplaces of husbands and wives	79
5.9	An evaluation of the wave of advance model	81
6.1	Map of the Rh− gene	88
6.2	Comparative simulations of two populations	91

6.3	Theoretical selection curves for the Rh gene	93
6.4	Map of the gene frequency of blood group A in Europe	95
6.5	Map of the gene frequency of blood group B in Europe	96
6.6	Map of the gene frequency of blood group O in Europe	97
6.7	Map of the frequency of the HLA-B8 allele in Europe	99
6.8	Map of the frequency of the HLA-BW15 allele in Europe	100
6.9	Illustration of a principal component	103
6.10	Contour map of the first principal component	105
6.11	Contour map of the second principal component	106
6.12	Contour map of the third principal component	107
7.1	Computer simulation of the wave of advance	110
7.2	Two models of movement between settlements	111
7.3	Examples of BANDK 2 simulation runs	114
7.4	Migration distributions from simulation run C	115
7.5	Example of the simulation of genetic patterns	122
7.6	The expansion of farming populations during the macro-simulation	124
7.7	Distribution of one gene from the macro-simulation	127
7.8	Example of a demic cline	128
7.9	Demic clines under different combinations of parameters	129
7.10	Example of the first principal component	131

TABLES

2.1	Plant remains present at early sites in the Near East	20
3.1	Composition of the faunal remains recovered from early neolithic sites in Europe	41
5.1	Rates of population growth and the time required for a population to double in size	71

PREFACE

The aim of this book is to bring together and provide an overview of the various studies that we have undertaken over the last twelve years on the shift from hunting and gathering to early farming as a new way of life in Europe and the implications of the neolithic transition for the genetic structure of human populations in Europe. One of the challenges that we have had to face in writing the book involves the range of academic backgrounds that different readers are likely to possess. On one hand, there is the need to present archaeological material to those whose training and experience are primarily in human genetics and the biological sciences. On the other, concepts and quantitative methods used in human population genetics must be developed in a clear manner for those coming from backgrounds in archaeology and the social sciences. We have thus tried to concentrate on the larger picture in the body of the text and to place more technical material and points of interest to specialists in notes to the respective chapters.

It is worth stressing at the outset that the chapters of the book dealing specifically with the neolithic transition are not written as a "prehistory" in the traditional sense of the term: that is, the fashioning of a narrative that attempts to portray diverse facets of the early life of a given region or period. Rather, we have adopted a more limited and thematic focus, essentially that of outlining what is currently known about the origins of early farming in Europe. Readers who become interested in more detailed regional studies may find it useful to consult the references listed in the bibliography.

Another point that needs to be made here is that our current knowledge of neolithic sites in most parts of Europe is still quite limited. Moreover, the nature of what is known is often dependent on when and how archaeological fieldwork was conducted. We can fully expect our understanding of the neolithic transition in Europe to increase substantially as further fieldwork is done over the next twenty-five years. Our own experience working at early neolithic sites in the region of Calabria in southern Italy, which is described elsewhere, offers a good example of how rapidly our knowledge of an area can change once intensive research is initiated. On a more technical note, radiocarbon dates are cited throughout the book in conventional C-14 years and not calibrated ones, since the calibration curves available at the present time (see Chapter 4) extend back only to about 7000 B.P. and do not cover the full time range of our study.

A few words should also be said regarding the hypothesis of demic diffusion that is developed in the book and the changing fashions in archaeological explanation. Diffusion was for many years one of the leading forms of explanation in archaeology. More recently archaeologists have come to be more cautious and selective in its use. One of the criticisms commonly leveled against diffusion is that not enough attention is paid to process. Geographical distributions of artifacts or other remains tend to become in themselves their own explanations. As part of the reaction against earlier diffusionism, a counter position—what for want of a better term might be called "indigenism"—has become popular in some quarters. Sources of change and innovation are sought within indigenous populations or local cultural contexts. This position, in which cultural change is cast as a self-contained affair, is probably no more tenable in its extreme form than the one it attempts to replace. Individual case studies need to be examined on their own terms and not forced into rigid molds. There may be less conflict between the study of diffusionary processes and the recent emphasis on endogenous cultural developments in prehistory than is often implied. Here we shall put forward the wave of advance model as a means of explaining a major part of the spread of early farming in Europe. As we shall see in the discussion of this model in Chapter 5, processes operating essentially at the local level will be employed to account for the diffusion of early farming in Europe.

The science of genetics is scarcely more than a century old, and yet it has given us a new dimension to the biological world. The past fifty years in particular have witnessed the rapid accumulation of a vast body of genetic data on human populations. The challenge to the geneticist remains to discover patterns in this wealth of information and to explain those patterns recognized. There has emerged in the last few years an increasing awareness of the complex relationships that may hold between genes and culture in human populations. It has been traditional in much of Western thought for the biological nature of groups or individuals to be regarded as determining cultural events or historical developments. In the case of the neolithic transition in Europe, it would appear

PREFACE

that the arrow of causality is reversed. We shall argue that cultural events in the remote past played a major role in shaping the genetic structure of human populations in this part of the world.

We would like to thank the many archaeologists and geneticists who over the years have either provided us with information or offered useful comments on our work. Special gratitude is extended to the following persons who collaborated in various aspects of the research done at Stanford University: Marcus Feldman, Juliana Hwang, Paolo Menozzi, Alberto Piazza, Sabina Rendine, Laura Sgaramella-Zonta, and Diane Wagener. Finally, we would like to acknowledge support that we have received from the National Institute of Health, the Department of Energy, and the National Science Foundation.

THE
NEOLITHIC TRANSITION
AND THE GENETICS
OF POPULATIONS IN
EUROPE

CHAPTER 1

INTRODUCTION

Human evolution is now thought to extend back well over two million years, but it is only during the last ten thousand years that food production has emerged as the main way by which humans provide for their subsistence. Previously, people met their food needs by means of hunting and gathering—that is, through the exploitation of seasonally available wild animals and plants. Agriculture, therefore, is a recent development when viewed in terms of the full course of human evolution. Associated with the shift to farming as a way of life are changes in technology, demography, and social organization. In Europe, this transformation did not occur as a sudden event but involved processes that required many generations to be worked out. Thus, rather than thinking in terms of a "neolithic revolution," as did the prehistorian V. Gordon Childe, we prefer to speak of a neolithic transition. In trying to account for the spread of early farming in Europe, we began to realize that the neolithic transition may have a major bearing on patterns of genetic variation observed among human populations in Europe. The interpretation of such patterns has long represented a challenge to population geneticists. There is the intriguing possibility that two seemingly unrelated research problems, one concerned with cultural development in the remote past and the other with the genetic structure of living populations, are closely linked.

At first glance, it would appear to be surprising for two researchers such as the authors, coming from fields as different from one another as prehistoric archaeology and population genetics, to find themselves working together. When we first began our collaboration in 1970, it was far from clear where the study would lead, and much of our initial effort was spent in trying to bridge the differences in our backgrounds. As various lines of investigation have unfolded over the last twelve years, it has been our experience that basic questions have become more clearly defined, and there is a better sense of how answers can be obtained. The aim of this book is to bring together the various studies that we have undertaken and to attempt a synthesis of the work that has been done in collaboration.

Very few populations in the world today live on the basis of hunting and gathering; almost all of the some five billion people composing the world's current population are sustained by foods derived from agriculture. If we could go back to 10,000 B.C., we would find that almost all populations had subsistence economies based solely on hunting and gathering (see figure 1.1). Moreover, rough estimates would place the world's total population at that time on the order of ten million people.[1] Our ideas about hunting and gathering as a way of life are drawn primarily from ethnographic studies of surviving hunter-gatherer groups, such as the bushmen and pygmies of Africa and the aborigines of Australia. One obvious limitation to such studies from a broad, comparative point of view is that hunter-gatherers tend to survive only in more extreme environmental settings or in those not suited to the practice of agriculture. Allowing for this possible bias, some common features of hunters and gatherers can be recognized, such as their small population sizes and their low population densities. Most hunter-gatherers exploit a wide variety of plant and animal foods and do not experience food shortages. Contrary to the view once held that hunter-gatherers lead a harsh life dominated by an endless quest for food, they appear able to meet their subsistence needs in most cases through the expenditure of only modest amounts of effort. It is common for hunters and gatherers to live in small camps that range in size from 15 to 40 people. Such a group or band, as anthropologists would refer to it, often follows a seasonal pattern of movement, occupying a series of different camps during the course of a year. The light and easily transported material culture that such groups usually possess would seem to be well adapted to a nomadic life style. One of the striking features of hunting and gathering, especially in contrast with modern ways of life, is the inherent ecological balance that this way of life entails. As anthropologists have come to take a more positive view of hunters and gatherers, the explanation of the origins of agriculture has become a more complex matter. In the next chapter, instead of regarding agriculture as an innovation providing a solution to recurrent food shortages among hunters and gatherers, we shall ask: Why did the shift to food production occur at all, and why did it take place in certain contexts and not others?

At archaeological sites such as Tell Aswad in the Near East, which dates to the eighth millennium B.C. (see Table 2.1), we first begin to see positive evidence for the cultivation of cereal crops. Since domesticated plants and animals are required for food production,

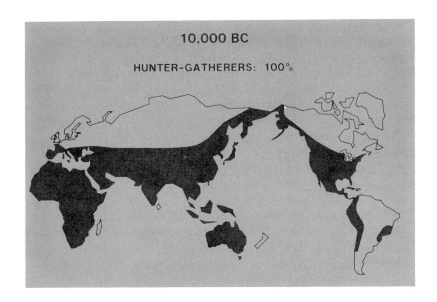

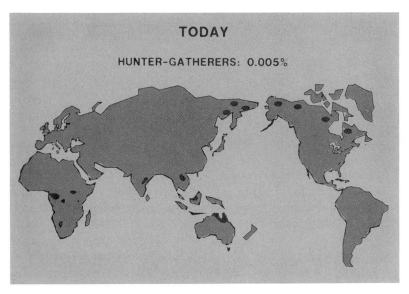

Figure 1.1. Distribution maps of hunters and gatherers at 10,000 B.C. and at the present time (after Lee and DeVore 1968).

the processes of domestication may have started at an even earlier date. The domestication of a plant or animal involves human interference in the reproduction of a population, leading to changes over time in its morphology and behavior. Traits considered to be favorable for purposes of food production, such as a more uniform time of ripening in domesticated strains of wheat, are obtained by means of artificial selection. Various aspects of plant and animal domestication and current explanations for the origins of agriculture in southwest Asia will be examined in Chapter 2. Well-developed systems of food production based on the cultivation of wheat and barley and the herding of sheep, goats, pigs, and cattle are documented by 6000 B.C. at archaeological sites that have a geographic distribution from Iran to Greece. With the exception of Greece, early farming sites are not observed in other parts of Europe at this date. Over the next two thousand years, agriculture as a way of life was established in most areas of Europe. The middle chapters of this book are devoted to the basic question of how this transition occurred. The early farming cultures that are found in different regions of Europe are introduced in Chapter 3. It is worth noting here that wheat and barley were not domesticated locally within various regions of Europe but were introduced into Europe from the Near East.[2] Insight into how this transfer occurred can be gained, as we shall see in Chapter 4, by measuring the rate of spread of early farming in Europe. This was the initial question that we explored at the beginning of our collaboration. The slow, regular pattern of the spread as a whole suggested a model for explaining the diffusion of early farming that is developed in Chapter 5. The wave of advance model, as it is called, regards population growth and local migratory activity as setting up an expanding population wave that advances at a steady radial rate. The model has major implications for the patterns of genetic variation that would be established among populations in Europe, if such a population expansion did occur.

At the conceptual level, it is worth distinguishing two modes of explanation for the spread of early farming in Europe. The first of these involves *cultural* diffusion, by which cereals and farming techniques are passed from one local group to the next without geographic displacement of groups; the second is what we have called *demic* diffusion, where the spread is due to the movement of farmers themselves. These two modes of explanation are not necessarily mutually exclusive: the real question may well be to evaluate their relative importance in different regions of Europe.

INTRODUCTION

As mentioned in the preface, archaeologists have become more cautious in their use of diffusionary explanations during the last twenty years, in part as a reaction against the abuses of diffusionism in the earlier literature. One of the main problems was that the distribution of a given object or trait was often considered to explain in itself what was happening; no real explanation of the process or processes involved was actually put forward. In addition, diffusion seemed to imply that cultural change derives in large measure from external factors. More recent literature places increasing emphasis on explaining cultural change in terms of endogenous factors or developments taking place within a society or population.[3] Yet, although the reaction against diffusionism has been a positive one in many respects, there has also been some failure to realize that diffusionary processes can occur in some cases as a result of events essentially at the local level. Moreover, adopting a regional perspective, as is frequently done in the more recent literature, opens the temptation to regard the region itself as an actor in a process, when it may only be the spatial framework or stage setting for the play. The actors are more likely to be people or cultures, and their drama as it is acted out over time may not always stay within the confines of their original theater.

As indicated above, the hypothesis of a demic expansion of early farming in Europe has major genetic implications. In Chapter 6 we turn to the subject of human genetics and the development of methods for characterizing genetic patterns among European populations. During the last forty years considerable strides have been made in understanding how the forces of evolution operate at the population level. At the same time, a vast body of data on gene frequencies has accumulated for human populations in Europe. The interpretation of maps of gene frequencies has nevertheless long remained a challenge to geneticists. When we look at maps of individual genes, such as those belonging to the Rh and ABO blood group systems, we usually see quite different patterns of variation from one map to the next. What are the factors that produce these differences? To what extent can we recognize common patterns for different genes? In order to answer the latter question, new methods for analyzing many different genes all at the same time had to be introduced, and considerable effort was spent to find methods appropriate for the generation of synthetic gene maps. Once such maps were obtained, they showed an interesting relationship with patterns expected under the demic hypothesis for the spread of early farming in Europe. From the viewpoint of the geneticist, order began to emerge from what had been a chaotic and uninterpretable accumulation of genetic facts. The underlying patterns were apparently connected with major cultural developments in the past that had left a deep imprint on the genetic structure of populations, and the patterns persisted during the course of those populations' subsequent histories.

As one way of evaluating whether the approach we adopted to the analysis of gene maps is an appropriate one, simulation studies were undertaken that involved populations in Europe experiencing first genetic differentiation and then, with the introduction of agriculture in southwest Asia, a demic expansion. This work, described in Chapter 7, provides support for the methods employed in generating the synthetic gene maps. From a heuristic point of view, the construction and operation of a simulation model also provide a means of exploring the internal logic of a process. In the present case, the simulation studies are particularly useful in drawing attention to the important role played by interactions between farming and hunter-gatherer populations during the spread of early farming and in showing how genetic change at the population level can occur in the context of cultural events.

CHAPTER 2

THE ORIGINS OF AGRICULTURE

2.1 Introduction

The shift from hunting and gathering as a way of life to a reliance upon food production represents one of the major transformations in the course of human evolution. Although the origins of agriculture have long been a subject for inquiry, it is only during the last fifty years that archaeological investigations have begun to provide direct lines of evidence. For example, we now have a reasonably clear idea about when and where domesticated forms of cereals such as wheat and barley made their appearance in southwest Asia. At the same time, a fair amount of evidence is now available on the cultural context in which these early crops were domesticated. It is of equal interest, as we shall see later in this chapter, to consider how early agriculture developed as an economic system.

Familiarity with our own agricultural and industrial ways of life makes it difficult for us to appreciate the scope of the transformation brought about by food production or to imagine what life would be like without agriculture. Studies of contemporary groups of hunters and gatherers such as the bushmen and pygmies in Africa and the aborigines of Australia provide some insight into this way of life. As mentioned in Chapter 1, hunter-gatherers usually live in small bands that shift their camps at various times during the course of a year. For food, they exploit a wide range of seasonally available wild plants and animals. Recent studies indicate that, rather than having to expend a good deal of time and energy in fulfilling their dietary needs, hunters and gatherers in most cases can sustain themselves with only a modest work effort and that they actually have a fair amount of leisure time. Food is not commonly stored for future use but is usually consumed within a few days of its collection. Low population densities, which, except among those exploiting fish, seldom reach the level of one person per square kilometer, appear to be a common feature of this way of life. As mentioned before, a rough estimate of ten million people has been put forward for the total world population of hunters and gatherers at 10,000 B.C., the period just prior to the origins of agriculture. This figure compares with a current world population of some five billion people.

Concentration on the production of one or only a few crops seems to be a regular feature of most of the farming systems practiced in the world today. Three cereal crops—wheat, rice, and maize—account for a combined annual production of almost one billion metric tons, providing the bulk of the food consumed by the world's current population. Properties shared by these three cereals are their comparatively high caloric values as foods and their high yields as crops. Even in cases where subsistence forms of farming are still practiced, yields are usually well over half a ton per hectare of land. Such yields are by no means a recent development; estimates on the order of one ton per hectare have been made for wheat yields in Mesopotamia during Sumerian times as well as for ancient Greece and Rome. In situations where more modern or intensive techniques of farming are employed, yields of several tons per hectare can be obtained. One implication of such yields is that a substantial proportion of a person's annual food requirements can be met today through the cultivation of only a fraction of a hectare of land. Another is that high local population densities can readily be sustained in this way. As a consequence of agriculture, new relationships have been established between the land, its use as a resource, and human populations.

Over the last one hundred years, scientists from a wide range of disciplines—agronomy, genetics, geography, and anthropology, to name only a few—have contributed to the study of the origins of agriculture. In order to gain some sense of historical development, it is worth recalling the contributions of a few of the main workers, starting with the Swiss botanist Alphonse de Candolle. In his treatise on the *Origins of Cultivated Plants* (1884), de Candolle attempted to locate the region of origin of some two hundred cultivated plants by drawing upon information from a wide variety of sources. Considering the rather limited nature of the evidence at the time, especially that available from dated deposits at archaeological sites, he performed an admirable, if highly speculative, intellectual exercise.

One of the concepts used by de Candolle, that of centers of origin for cultivated plants, was subsequently taken up and developed by N. I. Vavilov, a Russian agronomist and geneticist, who argued that there should be a close relationship between centers of genetic variation and centers of domestication. In 1926 Vavilov put forward his theory that the center of origin of a crop could be determined

INTRODUCTION   11

by analyzing the patterns of variation observed among living plant populations. The region in which one found the greatest genetic diversity of a crop plant was its region of origin, especially if the region also contained wild races of the crop in question. In this way, eight main centers were identified in different parts of the world. The work of Vavilov and his colleagues produced much useful information on crop plants, but their approach also had its limitations. Some of the crops examined do not seem to have clear centers of diversity, and not all crops appear to have been domesticated in Vavilovian centers. Rather than reflecting the place at which a plant was domesticated, modern patterns of genetic variation may be highly influenced by a crop's subsequent history of cultivation.

De Candolle and Vavilov were primarily interested in the geography of crop plants; neither tried to develop a theory about how agriculture originated as such. Of the many early explanations proposed for the origins of agriculture, perhaps the one attracting the widest attention was a theory first put forward in the late 1920s by V. Gordon Childe, whose main field of interest was social and economic development in prehistoric times. His propinquity theory was based on the notion that climatic change in the period just following the end of the last Ice Age resulted in warmer and drier conditions over much of North Africa and southwest Asia. Progressive dessication forced groups to move to areas along perennial rivers and near oases. The buildup of people in those relatively restricted areas where water was available created a need to increase the local food supply and, in turn, to domesticate plants and animals. Although the oasis theory is no longer accepted for reasons to be discussed later in this chapter, it nevertheless represented an attempt to explain in economic terms why the process of domestication was initiated. It was perhaps of equal importance in stimulating an interest in the testing of this and other theories in the field.

Archaeological investigations have come to play a central role in the study of the origins of agriculture only since the Second World War. Much of the credit for this new development goes to Robert J. Braidwood at the University of Chicago, who conducted several major field projects in southwest Asia. He took the position that the place to look for evidence on the domestication of wheat, barley, and sheep would be within what he called the "natural habitat zone," or those areas where wild progenitors of the respective species lived. He initially identified this zone to be the hilly region running in an arc along the interior of the Taurus-Zagros chain of mountains. The nuclear zone was later enlarged by Braidwood to include the southern flanks of the high Anatolian Plateau and the hilly regions of Lebanon and Judea. Climatic change was not regarded as playing a significant role in the origins of agriculture; attention was directed instead toward cultural factors. Where Braidwood made his main contribution, however, was in demonstrating the kinds of evidence that could be obtained in the field. The recovery of animal bones and carbonized seeds from early sites meant that discussions about plant and animal domestication could be placed on a more empirical footing. Archaeological excavations provided at the same time a means of reconstructing the cultural and environmental setting in which domestication took place. Such research involves the development of many different lines of evidence, and Braidwood was among the first in southwest Asia to realize that what was needed was a team approach, with archaeologists working in collaboration with specialists from other disciplines.

A distinction that deserves to be made here, one that is suggested by contrasting the approach of de Candolle and Vavilov on the one hand with that of Childe and Braidwood on the other, is that the origins of agriculture actually involve two related but somewhat different stories. One takes as its central theme the question of when and where various plants and animals were brought under domestication. The narrative unfolds by turning from one species and the evidence for its domestication to the next. In this case, what is understood by the term "origin" is essentially the history of domestication. The other story is more concerned with the emergence of agriculture as an economic system. Attention is focused on the economic motives for domestication and the role that a domesticated plant or animal plays within a system of food production. There is an interest in the integration of several different plants and animals into a comprehensive system for meeting subsistence needs. Origins here refer to the formation of early subsistence economies oriented toward food production.

Some care must be taken so that one story is not confused with the other. For instance, the appearance of a domesticated plant or animal is not necessarily equivalent to the presence of an economic system based on food production. As in the case of maize in Mesoamerica, there may be some time between initial efforts at the cultivation of a plant and eventual reliance upon it as a major crop. Likewise, plants that are brought under domestication within the

MULTIPLE CENTERS OF ORIGIN

FIGURE 2.1. Centers of agricultural origins in the world. Harlan (1971: fig. 6) identifies three reasonably compact centers (A1, B1, and C1) in the Northern Hemisphere and three broader areas (so-called non-centers: A2, B2, and C2) in the Southern Hemisphere.

context of agricultural systems that have been operating for some time, such as grapes and olives in the eastern Mediterranean, should be regarded as contributing to the elaboration and growth of existing systems of food production and not to the actual origins of such systems. This distinction is of particular relevance when it comes to the study of early agricultural systems in Europe, as we shall see in the next chapter. It is worth stressing that the two stories are in many respects complementary to one another and that both need to be studied.

2.2 MULTIPLE CENTERS OF ORIGIN

Current evidence suggests that agriculture arose as an independent development in several different parts of the world. Three main centers are shown on a map of agricultural origins prepared by J. R. Harlan in the early 1970s (see figure 2.1): one in southwest Asia for wheat and barley; another in north China, where foxtail millet appears to have been the main crop initially; and a third one in Mesoamerica, where maize and several other plants were cultivated at an early date. Harlan also proposes three larger zones in the tropics, which he calls "non-centers," where a variety of other plants were presumably domesticated at early dates. One positive aspect of the concept of non-centers is that it recognizes the greater ecological diversity of tropical zones. It also allows for the possibility that somewhat different processes of domestication may operate in such zones. Although some supporting paleobotanical evidence is starting to become available, it is still too early to document in any detail the configurations of the non-centers or to evaluate the utility of the concept.

The center with the earliest evidence for domesticated cereals is the one in southwest Asia, which essentially corresponds with the nuclear zone proposed by Braidwood.[1] The remains of both wild and domesticated forms of wheat and barley are observed at sites occupied during the period between roughly 8000 B.C. and 7000 B.C. As we shall see below, efforts to bring these crops under cultivation were made in the context of permanently occupied villages. There is good evidence for the cultivation of wheat, barley, peas, and lentils by 6500 B.C. in various parts of southwest Asia. By 6000 B.C., well-developed systems of food production had emerged in which animal husbandry was fully integrated with the cultivation of crops.

The second of Harlan's centers is located in north China, where Shensi Province contains a large number of villages belonging to the Yang-shao culture. The subsistence economies of these villages, which date to the period between 4500 B.C. and 3500 B.C., are based on domesticated foxtail millet and pig as well as on a variety of wild plant foods, including chestnuts, hazelnuts, and pine seed. The antecedents of the Yang-shao culture document that food production in this part of China extends back at least into the sixth millennium B.C. Much less is known at present about when and where rice was domesticated. This process may have occurred in one or more places over the wide area extending from eastern India to southern China. There is evidence for the cultivation of rice at the site of Hemudu in south China that dates to the late sixth millennium B.C.[2] In Thailand, impressions of rice grains are seen on pottery that dates to about 4000 B.C. Rice is not observed, however, in either wild or domesticated form among the plant remains from the Spirit Cave in Thailand, which dates back to about 7000 B.C. Clarification of the situation in southeast Asia can be expected as further work on early agriculture is conducted in the region over the next decade.

The third center recognized by Harlan, the one in Mesoamerica, is located roughly between Mexico City and Honduras. Excavations by R. S. MacNeish and his colleagues in the Tehuacan Valley provide evidence for some of the earliest cultivated plants in Mesoamerica. There is still some uncertainty about which crop plants are the earliest in this region; claims have been advanced for the grain *Setaria* and for the chili pepper in the time range between 7000 B.C. and 6000 B.C. Maize itself is documented in the Tehuacan Valley by 5700 B.C. The exploitation of this crop in Mesoamerica may well have started at even an earlier date. There is still active debate among specialists over the identity of the wild progenitor of maize. The two rival candidates are the crop weed teosinte (*Zea mexicana*) and a wild form of maize that subsequently became extinct. The earliest corn cobs are very small, and over time there is a definite trend toward increasing cob size. It has been suggested that the yields of early maize fields were quite low, perhaps only on the order of 60-80 kilograms per hectare during the early stages of domestication.[3] In contrast with the situation in southwest Asia, the domestication of maize took place in the context of seasonally migrating groups rather than sedentary villages. It is only in the so-called Ajalpan phase of occupation, starting about 1500 B.C., that full-time farmers living in villages are recognized in the Tehuacan Valley.

Early evidence for plant domestication is also observed in South America. Domesticated forms of two species of beans and also chili pepper are apparently present in the Andean Highlands by 6500 B.C. It is worth commenting that maize seems to make its appearance in the Andean area only at a date of about 3000 B.C., which would indicate a considerable time lag with respect to its initial occurrence in Mesoamerica.

Sub-Saharan Africa represents another part of the world where local forms of agriculture were independently developed. It is widely held that indigenous plants were brought under cultivation within a broad band running across the width of the continent below the Sahara. Among the main cultivated plants here are sorghum, pearl millet, finger millet, African rice, and yams. One of the distinctive features of African agriculture even today is the highly localized geographic distribution of many cultivated plants. Although there is still relatively little evidence in the form of plant remains from archaeological sites, the work that has been done would seem to indicate that most of the plants came under cultivation only after 3500 B.C.[4] It is possible that events in the Sahara may have contributed to agricultural origins in sub-Saharan Africa. Desmond

Clark has noted that the onset of desiccation in the Sahara at about 4000 B.C. (for which there is much better evidence than the early postglacial desiccation in North Africa hypothesized by Childe) would have made it increasingly difficult for pastoralists to live as they previously had done in the Sahara. There is now evidence that cattle pastoralism was well established in some parts of the Sahara by 6000 B.C.[5] The movement of pastoralists into the savanna zone south of the Sahara would have promoted a local imbalance in resource exploitation, which may have stimulated resident populations to intensify experimentation with indigenous plants.

A review of the evidence related to the origins of agriculture on a worldwide scale shows diversity to be one of the most striking features. With few exceptions, the plants initially brought under cultivation are not the same crops but differ from one part of the world to the next. There are also major differences in terms of the ecological and cultural settings in which various crops are brought under domestication. One of the few variables that would seem to be shared is timing: early experiments at plant domestication occurred in southwest Asia, east Asia, and Central America during the period between 8000 B.C. and 5500 B.C. Such a coincidence raises the question of why developments that are presumably independent from one another should be taking place in distant parts of the world at roughly the same time. This intriguing question cannot be answered at the present time. M. N. Cohen has suggested that population pressure, seen as a widespread phenomenon during the late Pleistocene and early Holocene, may be responsible for these developments. The archaeological evidence currently available would seem to provide only limited support for this explanation. As more evidence accumulates about the pathways to agriculture in different parts of the world, we can expect to be in a better position for evaluating whether some general factor or process is involved or whether the appearance of multiple centers of origin must be explained in terms of different combinations of local factors.[6]

2.3 Plant Domestication

Domestication can be viewed as an evolutionary process under human influence. Through breeding or the manipulation of the genetic structure of a population, long-term changes are obtained in a plant or animal. Modification does not occur as a sudden event but builds up gradually and cumulatively over the course of generations. It is important to remember that although we have acquired an understanding of the genetic basis for inheritance only during the last one hundred years, the successful breeding of plants and animals—not as a modern breeding program but as a practical matter of everyday life—has a long history. The primary aim of husbandry has traditionally been the pragmatic one of trying to establish those morphological and biological traits within a population that human groups consider to be desirable.

It is worth mentioning certain traits that might be considered beneficial in the case of cereal crops. These would include both greater seed size and a larger number of seeds per plant. Morphological changes such as the loss of glumes in barley would make food preparation a much easier task. Among wild plants, there is often considerable variability in the time that individual plants reach maturity; selection for a more uniform maturation time can make harvesting easier and more productive. There are also advantages in developing strains that permit wider environmental tolerance or greater resistance to disease. Such beneficial changes are not achieved without a cost to the plant, however. A species that is fully domesticated may become to a large extent dependent upon humans for its survival. This is particularly true for plants that lose their natural mechanisms for seed dispersal and come to rely on sowing or planting by humans for their propagation. A cultivated field is not a natural habitat: human activities—cultivation, sowing of seed, weeding, and so forth—are required to maintain such an environment. At the same time, it has been said that in making such a commitment of time and energy to the well-being of plants, human beings have in a sense domesticated themselves.

The key step in the domestication of a cereal such as wheat involves the intentional sowing of seeds. By itself, the intensive harvesting of a wild plant tends to have only limited effects on the genetic structure of a population. It is only with the sowing or planting of harvested seed that certain major processes of selection are set in motion. One of the principal changes in crops such as wheat and barley is the establishment of the so-called nonshattering trait. Shattering in wild cereals provides the means by which seed dispersal is accomplished. In wild forms of wheat, for example, the ear consists of several spikelets connected by rachis segments (see figure 2.2), which are described as "brittle" because they shatter easily when the grain is ripe. The ear will hold together better if there is selection for a nonbrittle rachis. This modification is automatically selected for during harvesting, since those seeds that

FIGURE 2.2. Diagram of a cereal spike. The black column is known as the rachis. Its component parts, the internodes, are separated by white interstices denoting the points at which, in a wild cereal, the rachis shatters into sections consisting of the internode with one or more attached spikelets of grain. Under domestication, these disintegration points become solid, and a nonbrittle rachis is formed (Helbaek 1969: fig. 138).

do not shatter tend to be recovered in greater number than those that do. As the process of harvesting and sowing is repeated year after year, the nonshattering trait becomes established in a population. The crop as a whole thus becomes easier to harvest and yet more dependent upon humans for its propagation.

That wheat and barley are both self-pollinating plants was undoubtedly a significant factor in their early domestication. In contrast with cross-pollination, which is found in crops such as maize, where it may have contributed to a much slower rate of domestication, self-pollination means that fertilization normally occurs within the same plant and allows a population to be split in effect into independent pure lines. Under this kind of system, new forms, such as cultivated wheat with a nonbrittle rachis and wild wheat with a brittle rachis, can coexist in the same locality without having to face the prospect that the new type will be swamped by the more common (wild) one because of cross-pollination from the wild wheat. In a sense, the reproductive system of self-pollinating plants provides protection for modifications that begin to emerge during early stages of the domestication process.

Another factor contributing to the development of early crops would have been the movement of seeds from one area to another. Differentiation between plant populations can occur when seeds are transported to a new, geographically isolated area. On the other hand, movement can also bring seeds from distant areas together with local ones, thus raising the possibility of hybridization and the incorporation of new germ plasm in local populations. Cycles of differentiation and hybridization no doubt played major roles in the early evolution of wheat and barley as crops.

In classifying the various species of wheat, three main groups can be distinguished on the basis of their levels of polyploidy, or the number of chromosomes that a plant possesses. The basic number of chromosomes in the wheat genome is seven, and the three levels of the polyploid series are, respectively, diploid ($2n = 14$), tetraploid ($4n = 28$), and hexaploid ($6n = 42$). Forms at a given level can hybridize readily with one another. Four species of wheat are of particular interest as far as early domestication is concerned: wild einkorn (*Triticum boeoticum*), a diploid, which gives rise to domesticated einkorn (*Triticum monococcum*), and wild emmer (*Triticum dicoccoides*), a tetraploid, which gives rise to domesticated emmer (*Triticum dicoccum*). A fifth species of interest is bread wheat (*Triticum aestivum*), a domesticated hexaploid form that has no known wild hexaploid progenitor. On the basis of cytogenetic research,

TABLE 2.1 Plant remains present at early sites in the Near East

Site	DATE B.C.	Wild einkorn	Einkorn	Wild emmer	Emmer	Wild 2-row barley	Hull 2-row barley	"Naked" barley	Pea	Lentil	Pistachio
Abu Hureyra	8750	+	.	.	.	+	.	.	.	+	.
Mureybat	8000	+	.	.	.	+	.	.	.	+	+
Tell Aswad	7750	.	.	.	+	+	?	.	+	+	+
Jericho	7500	.	+	.	+	.	+	.	+	+	.
Ali Kosh	7500	+	+	.	+	+	.	+	.	.	+
Çayönü	7250	+	+	+	+	+	.	.	+	.	+
Beidha	6750	.	.	.	+	+	.	.	.	.	+
Haçilar	6750	+	.	.	+	+	.	+	.	+	.

Sources: Abu Hureyra (Moore 1975; Moore 1979); Tell Aswad Ia (Van Zeist and Bakker-Heeres 1979); all other sites (Van Zeist 1976: Table 1).

Note: The dates are based upon uncalibrated radiocarbon dates (see Site Lists 4.1 and 4.2) and are cited in terms of the nearest quarter millennium.

bread wheat, which is documented in the archaeological record by 6000 B.C., seems to have originated through the hybridization of a domesticated tetraploid wheat and the diploid wheat *Aegilops squarrosa*.

The two main lines of paleobotanical evidence that can be used for studying early crops are carbonized seeds and seed impressions left in clay. Domesticated forms of wheat or barley can be distinguished from wild forms on the basis of the morphology of seeds and rachis fragments. The occurrence of plant remains at some of the early sites in southwest Asia is summarized in table 2.1. Domesticated forms of einkorn, emmer, and two-row barley all make their appearance by 7000 B.C. It is worth noting that several different cereal and legume crops are often observed together at a given site. Knowledge of the crops occurring at archaeological sites makes it possible not only to date their appearance as domesticates but also to ask questions about the ecology and geographic distribution of early crop plants. It is of considerable interest to know, for example, whether the distributions of the various wild progenitors of wheat and barley overlapped or were distinct from one another. Reconstructions of the geographic distributions of the wild progenitors have been put forward largely on the basis of modern field studies, which indicate that the natural habitats for wild wheat and barley are hilly areas with a sub-Mediterranean oak-park forest

ANIMAL DOMESTICATION

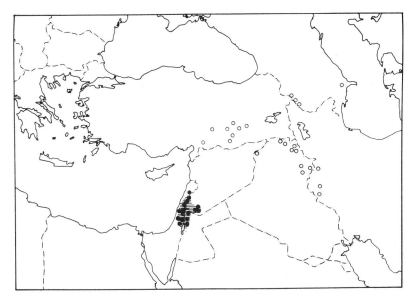

FIGURE 2.3. Distribution map of wild emmer in the Near East. Sites where wild emmer (*Triticum dicoccoides*) can be observed today in the Levant are indicated by solid dots. Hollow dots represent known sites of wild tetraploid wheats in Turkey, Iraq, and Iran (Zohary 1969: fig. 2).

type of vegetation. So far, one limitation of the reconstructed distributions is that they do not seem to take into account the climatic changes (see below) that occurred during the period between 10,000 B.C. and 7000 B.C. The wild cereals appear to have an affinity for the heavy soils found in areas with hard limestone or basalt bedrock. Wild einkorn, which in modern populations exhibits a range of morphological variations and ecotype adaptations, has a wide distribution extending along most of the Taurus-Zagros arc. Wild emmer, which has a more restricted distribution, is found locally in the Levant (see figure 2.3). Wild barley (*Hordeum spontaneum*) appears to have the widest geographic distribution of the three wild progenitors, extending in a full arc from the Jordan Valley to southwest Iran.[7]

2.4 ANIMAL DOMESTICATION

The domestication of several species of animals—sheep, goat, and pig—took place in southwest Asia at roughly the same time as the domestication of early cereal crops. Animal domestication presents some special problems, especially with regard to the documentation of behavioral adaptations and the actual identification of domesticated forms. In comparison with plant domestication, much smaller populations are involved, and once it is in practice, animal husbandry offers wide potential for artificial selection. Only a few individuals in a population may be responsible for the bulk of the offspring allowed to reach reproductive age in the next generation.

Much of the selection in early animal domestication was probably directed toward the modification of behavioral traits rather than morphological ones. Animals are in many respects more demanding than plants, owing to their mobility and their need for a regular food supply. This is particularly true for animals that exhibit seasonal patterns of movement, such as wild sheep and goats. There are two main adaptive strategies that can be employed here: either human groups can adopt a pattern of movement similar to that of the species to be brought under domestication, or the seasonal behavior of the animal can be modified, with human groups taking responsibility for seeing that a food supply is available throughout the year. There is a good chance that both strategies were explored in southwest Asia at an early date.

One of the main problems in the study of animal domestication concerns the criteria used in deciding whether a given animal is domesticated or not. Part of the difficulty stems from the fact that morphological changes within animal populations can develop slowly and that individual animal bones may not adequately reflect those changes that do occur. Even though large quantities of animal bones are often recovered during archaeological excavations, most bones do not permit a clear distinction to be made on morphological grounds between wild and domesticated forms. Those few bones that are more diagnostic, such as twisted horn cores in domesticated goats, usually occur in numbers that are too small to permit a statistical analysis of ranges of variation. These limitations in terms of morphology have led to the use of two other criteria for identifying domestication. One of these is based on the appearance in a region of a new species that has not previously been seen in the faunal record of that region. The inference is made that the new animal has been introduced in domesticated form, with the actual process of domestication occurring someplace other than in the region itself. An argument of this kind has been made for sheep in the Levant. In most cases, however, wild forms of sheep, goat, pig, and cattle have long histories in most parts of southwest Asia. This means of recognizing domestication is clearly of limited use in this region, although it has much wider application with respect to sheep and goats in Europe, as we shall see in the next chapter.

The other commonly used criterion of domestication is based on the age structure of a set of faunal remains. The age at which an animal is killed can be determined by studying certain bones, such as its teeth. Through the careful analysis of faunal remains, it is possible to reconstruct the age structure of the animals slaughtered at a site. In the case of hunted animals or a wild population, there is a good chance that the age structure will reflect a full range of age classes. On the other hand, it is common in the husbandry of domesticated animals, especially in areas where water and pasture are in short supply during the summer months, to slaughter a fair proportion of the young or juvenile animals. Animals that are allowed to reach maturity are not usually slaughtered until they have passed through their reproductive years. The age composition of faunal remains in this case will show many young animals and some older ones but relatively few animals of reproductive age. This is the pattern seen among the faunal remains of sheep from the site of Zawi Chemi Shanidar in Iraq, which dates to about 9000 B.C.

Some questions can be raised about the interpretation of patterns of this kind. It is known from field studies that the age structure among wild populations such as wild mountain sheep in western Canada can vary considerably from one population to the next.[8] There is also some evidence that hunters in certain cases are selective in the way in which they exploit a wild herd. Although the specific claim for domestication is weakened by such considerations, the repetition of this pattern at a number of sites in a region would at least imply a pre-adaptation to domestication in the sense that a herd structure similar to the one expected among domesticated populations is being maintained. Some authors have even suggested that a loose form of herding and population control may have been practiced by hunters for a substantial period of time prior to actual domestication. There is as yet little direct support for this idea, and further work in the field in southwest Asia will be required before it can be fully evaluated.

In terms of the evidence currently available, the earliest animal to be domesticated in southwest Asia would appear to be the dog at about 10,000 B.C. At several sites in the region dating to the period between 9000 B.C. and 7000 B.C., the remains of sheep and goats display an age structure indicative of domestication. By 7000 B.C. there is more direct morphological evidence for the domestication of sheep, goat, and pig at sites in southwest Asia.[9] It is worth adding that at any one site the initial efforts at domestication seem to be directed in most cases toward a single animal.

2.5 Explaining the Origins of Agriculture

The study of modern hunters and gatherers has changed our perspective on the economic developments involved in the origins of agriculture. Such studies offer little support for the view once widely held that groups of hunters and gatherers are engaged in a continual quest for food and regularly face the specter of starvation. Agriculture, in this view, could be explained as an innovation that arose in response to food problems. Apparently only a lack of knowledge about plants and animals delayed their domestication.

As mentioned earlier, however, it now appears that hunters and gatherers are not usually confronted with food shortages and can gain their subsistence in most cases without great difficulty. They often possess, in addition, a detailed and intimate knowledge of the habitats and seasonal behavior of plants and animals.[10] Ironically, this more positive view of hunting and gathering as a way of life has made it more difficult to account for departures from the mode of subsistence that has been common throughout most of human evolution. The question of why there should be a shift to food production has become more complex. At the same time, it is worth recalling that the economic systems of modern hunting and gathering populations show considerable variation in their food preferences, the techniques used for processing food, and the intensity of resource exploitation. One implication of this diversity is that it is probably unsound to use modern groups as direct guides to the adaptations leading to the emergence of food production. Just in terms of geographic distribution, surviving groups of hunters and gatherers are confined for the most part to desert, rain forest, and subarctic zones or areas of the world that are not suited to cereal agriculture.

The Levant, Iraqi Kurdistan, and the Nile Valley represent three areas where a fair amount of work has been done on the subsistence economies of pre-agricultural groups. In the Nile Valley of Upper Egypt and Nubia, extensive field work has been conducted by F. Wendorf and others on sites dating to the period between 16,000 B.C. and 10,000 B.C. The main animals hunted at this time were wild cattle, hartebeest, and gazelle. There is also evidence that fish were exploited as a food resource at some of the sites along the edges of temporary lakes filled by the flooding of the Nile. Although

EXPLAINING THE ORIGINS OF AGRICULTURE

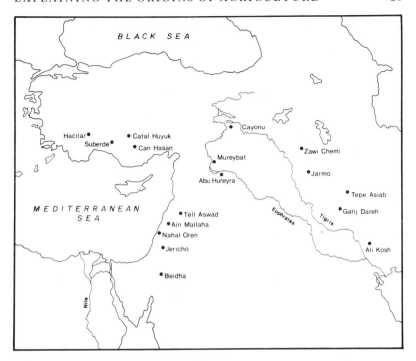

FIGURE 2.4. Sites in the Near East that date to the period 10,000–6000 B.C.

grinding stones presumably used for preparing wild cereals are regularly found at early sites along the Nile Valley, there is as yet no positive evidence in the form of plant remains for the exploitation of cereals (see note 1 of this chapter). Comparatively few sites are known for the period between 10,000 B.C. and 6000 B.C., and it is still unknown whether wild cereals were exploited during this time and whether they were actually domesticated in the Nile Valley.

Our knowledge of the situation in Iraqi Kurdistan comes largely from the excavations directed by Ralph Solecki at the Shanidar Cave and the nearby open settlement of Zawi Chemi. As indicated by evidence from the cave site, subsistence economies prior to 10,000 B.C. were based on the hunting of a range of wild animals and the collection of land snails. Zawi Chemi, which was occupied during the period roughly between 10,000 B.C. and 9000 B.C., documents a new kind of settlement in the area. It covers an area measuring some 200 meters in diameter and shows several phases of occupation. In the lowest levels, red deer is the most abundant species among a range of wild animals that are still being hunted. In the upper levels, the remains of sheep constitute almost 80% of the faunal assemblage and present an age structure considered to be characteristic of a domesticated population. A large number of grinding stones and querns were recovered during the course of the excavations at Zawi Chemi. The remains of circular-shaped house foundations made out of stones and the occurrence of many storage pits suggests a more permanent or sedentary mode of site occupation.

The best-known region of southwest Asia is the Levant, where a fair number of sites belonging to the Natufian culture and dating to the period between 10,000 B.C. and 8250 B.C. have been examined. Natufian settlements, which can encompass an area up to 7000 square meters, tend to be much larger in size than sites dating to the preceding Kebaran period. Architectural remains such as storage pits and round and oval-shaped houses are commonly found at Natufian sites. The stone tool inventory includes flint sickle blades and mortars and grinding stones that are often made out of blocks of basalt. Although paleobotanical evidence is still scarce, it would appear that wild cereals were exploited by the inhabitants of the Natufian sites. Fallow deer, gazelle, wild goat, wild boar, and wild cattle have all been recovered among the faunal remains. With the exception of the dog, there seems to be no clear evidence for animal domestication. Subsistence economies based on hunting and gathering apparently developed to the point at Natufian sites where a sedentary way of life was possible. The Natufian economic system was evidently viable enough to permit an expansion into areas of the Levant with less favorable environments such as the Negev.

In the northern part of the Levant the sites of Mureybat and Abu Hureyra, both of which date to the ninth millennium B.C., represent further examples of sedentary villages that developed within the context of hunting and gathering economies. Both sites occur as tells or large mounds that cover an area of at least two hectares and consist of a series of superimposed occupation layers. Grinding equipment and structural features including houses and storage pits are again regularly found at the two sites. The main sources of meat appear to have been wild cattle, wild ass, and gazelle. Some evidence for the exploitation of aquatic resources is seen at Abu Hureyra but apparently not at Mureybat, though both sites are located near the Euphrates River. Perhaps the most interesting finds relating to the subsistence economies of the sites are carbonized seeds of wild einkorn and wild barley, which are found

EXPLAINING THE ORIGINS OF AGRICULTURE 27

in some quantity. Yet the area in which the sites are located is considered by some specialists to be too dry to fall within the natural distributions of the wild cereals. Thus the inhabitants of the sites may well have been responsible for introducing wild cereals into the area along the Euphrates.

The general picture that emerges from this review of pre-agricultural sites is that similar kinds of adaptations seem to have occurred in different parts of southwest Asia at more or less the same time. Among the trends that can be observed are: (a) grinding equipment seems to have been used more frequently; (b) storage pits are found more commonly at sites; (c) the remains of buildings or houses are regularly encountered at sites; (d) settlements are larger in size; and (e) settlements appear to be occupied in a more permanent way. No clear explanation can be offered at the present time for why a sedentary form of village life arose during this period. Nevertheless, it is useful to consider some of the factors that may have been involved. One of these was the emergence of "broad spectrum" economies, as K. V. Flannery has called them, or economies in which a broader range of plants and animals is exploited. Resources such as small game, land snails, crabs, mussels, wild pulses, and wild cereals are harvested more actively and there is less reliance than previously on the hunting of larger herd mammals. This development would lead over the long term to a situation where certain groups would have a resource base diversified enough so that they could live in one place without having to relocate seasonally. The harvesting of wild cereals in particular may have close links with the emergence of sedentism. Stands of wild wheat and barley, if they are systematically harvested, can produce large amounts of grain, as is well illustrated by an experiment at harvesting a wild wheat field that Harlan conducted in Turkey. Using only a "prehistoric" sickle consisting of flint blades set in a wooden haft, Harlan found that he was able to harvest by hand the equivalent of about one kilogram of clean grain per hour. In the area of Turkey where Harlan made his experiment, there are extensive stands of wild einkorn that ripen at somewhat different times depending on the altitude of the area. He estimated that a family of four working every day for a period of three weeks following the ripening wild stands could harvest about one metric ton of wheat. Such a harvest would fulfill a major part of the annual food requirements of the family.

One prerequisite for consumption at later times would be the storage of much of the grain. Fortunately, cereals such as wheat and barley are well suited to long-term storage. The cereals are, however, more demanding when it comes to their preparation for consumption. Grinding or some other form of processing is needed to remove the tough, inedible glumes that adhere to the grain of wild wheat and wild barley. Grinding equipment facilitates the preparation of the wild cereals, and yet its possession places constraints on a group's mobility. We can thus view the exploitation of wild cereals, the possession of grinding equipment, the storage of grain, and sedentism as closely linked with one another in terms of functional relationships. In light of the apparent ease with which wild cereals can be harvested in quantity, the basic question shifts from how a sedentary way of village life could be maintained to why the cultivation of domesticated cereals should come to replace the exploitation of wild ones.

Before turning to this question, it is worth introducing another factor, climatic change, that may have influenced subsistence strategies and settlement patterns during the period between 10,000 B.C. and 7000 B.C. It is only during the last fifteen years that pollen studies have made it possible to resolve some of the longstanding debates about environmental change.[11] The identification of pollen grains extracted from samples taken at different depths in a pollen core permits the construction of diagrams that show the composition of local plant communities and how they change over time. The pollen diagrams are used in turn to make inferences about changes in climate.

Pollen diagrams obtained from cores taken from lakes in different parts of southwest Asia and the eastern Mediterranean reveal much the same general pattern. Prior to 9000 B.C. the vegetation was much more open and steppelike, indicating somewhat colder and drier climatic conditions at the time. During the period between 9000 B.C. and 7000 B.C., a change from steppe to open woodland vegetation occurred, reflecting a climatic regime that was both warmer and wetter and thus closer to a modern one. The results of the pollen studies do not agree with Childe's theory, nor do they support Braidwood's position. Significant changes in plant communities and probably also in the geographic distributions of many plant species took place during the period when plants and animals were beginning to be domesticated. However, it is still too early to evaluate the specific role that environmental change may have had either directly or indirectly in the adaptations leading to the origins of agriculture.

One of the suggestions prompted by the yields that can be ob-

EXPLAINING THE ORIGINS OF AGRICULTURE

tained by harvesting stands of wild cereals is that domestication may have taken place not within but outside of those areas where the stands naturally occur. An explanation for cereal domestication that incorporates this idea is the "density equilibrium" model put forward by Lewis R. Binford. He emphasizes the expected balance between human population density and the resources available to a population. Local population growth occurring within an area that is considered to be optimal in terms of the wild food resources it has to offer leads to a density where pressure on local resources begins to be felt. Stands of the wild cereals are assumed to occur in the optimal zone, and a sedentary way of life would itself be considered a major factor contributing to population growth. One of the ways for correcting such disequilibrium would be to have part of the local group move to adjacent marginal areas with low population densities. In order to improve the resource base in such marginal zones, which presumably lack stands of the wild cereals, those emigrating from an optimal zone would bring seeds with them. Planting the seeds would set in motion the domestication process. Though in many ways attractive, this model poses difficulties for testing in the field because archaeologists must determine whether a given early settlement was located in an optimal or a marginal zone.

Notwithstanding the high yields that can be obtained by harvesting stands of wild wheat and barley, there may be reasons for thinking that domestication could just as well have started within the natural habitat zones. There are probably more potential problems associated with wild grain harvests than recognized by Harlan in his experiment. The basic question here is not just one of yields but of the reliability of such harvests when a given stand of wild wheat or barley is repeatedly exploited year after year. In some cases one consequence of repeated, intensive exploitation may have been a decline in the quality and size of local stands. Factors such as pests and incursions of grazing animals may also affect the quality of stands in negative ways. One response to such deterioration may have been efforts toward the management of local stands, involving the sowing of seeds that were previously collected. In addition to trying to improve the density of stands and their sizes, efforts may have been made within the local context to encourage the growth of wild stands in new places. There might well be economic advantages, for example, in modifying the locations of wild stands so that they would correspond more closely with those of sites. Moreover, the kinds of sowing that would be called for in either case need be done only on an occasional basis.

As an account of how cereal domestication began in southwest Asia, the stand management model draws attention to the effects that repeated harvesting may have had on plant populations and how an adaptation such as sowing may have helped to correct such a situation. The management of wild stands is viewed as taking place essentially within local contexts and in response to fluctuations over time. The way in which sowing was done can also be expected to have varied somewhat from one place to the next. Under the model, the pace of artificial selection to be expected would have been quite slow initially, since sowing was probably not done on a regular basis. Distinguishing between wild and domesticated forms of the cereals is likely to be a demanding business for the paleobotanist or archaeologist during the initial stages of the domestication process, which may have lasted for a substantial period of time.

2.6 THE GROWTH OF FOOD PRODUCTION

Although the question of how plant and animal domestication began tends to dominate our thoughts on the origins of agriculture, the story of how several different crop plants and domesticated animals are integrated into an economic system and how such early systems of food production evolved is no less important. Only a brief overview of the development of early agricultural systems will be attempted in this section. In the case of sites such as Ali Kosh, stratigraphic sequences in a mound allow developments to be traced over a series of phases of occupation at the same site. As can be seen in Table 2.1, einkorn and barley that apparently occur only in wild forms are being exploited at Abu Hureyra and Mureybat in the ninth millennium B.C. Already by 7750 B.C. domesticated emmer and perhaps also barley are documented in the earliest phase of occupation at Tell Aswad. In a subsequent phase of occupation at the site, dating closer to 7000 B.C., domesticated forms of einkorn, durum wheat, and naked barley are also present. During the period between roughly 7500 B.C. and 6750 B.C., a wide geographic distribution can be observed for domesticated emmer in the Near East—Jericho, Ali Kosh, Çayönü, Beidha, and Haçilar, among the sites in Table 2.1. It is worth noting that two or more other cereals (often with barley occurring in a wild form) are seen at these sites, which suggests a certain diversity at each site in the

THE GROWTH OF FOOD PRODUCTION   31

exploitation of cereals. If we turn to sites dating closer to 6000 B.C., wild forms of wheat and barley are usually no longer seen, and three or more different domesticated cereals can commonly be observed. At Tell Ramad III in Syria, for example, three forms of domesticated barley, two of emmer, and einkorn have all been identified, as well as peas, lentils, vetch, and flax.

The evidence available at present suggests that animal domestication may have preceded that of plants. In addition to the possibility that sheep were domesticated at Zawi Chemi by 7500 B.C., domesticated goat is considered to be present at both Asiab and Ganj Dareh by that date. Thereafter, plants and animals were apparently developed as domesticates at the same sites. In the earliest levels at Ali Kosh, sheep and probably goats were herded. Domesticated goat and perhaps also pig are found at the site of Çayönü in Turkey. In the Levant, there is evidence for the domestication of pig at Jericho and possibly goat at Jericho and Beidha. The order in which animals are brought under domestication in southwest Asia appears to begin with the dog, followed by sheep and goat and then pig.[12] It is only at the comparatively late date of about 6000 B.C. that domesticated cattle make their appearance in Turkey. Animal domestication seems to differ from that of plants in an initial concentration on only one or at most two species at a given site, even though the faunal remains from early sites indicate no lack of other wild species in the local fauna. One suggestion here is that most of the requirements for meat and other animal products could be met initially by focusing on a single animal.

Turning to sites in southwest Asia dating to 6000 B.C., we can see that dry farming is well established and that three or four species of domesticated animals are represented at a given site. The herding of several different animals, each having its own set of behavioral characteristics, reflects the integration that has taken place within early systems of food production during the seventh millennium B.C. Economic growth is also apparent in the development of individual sites such as Ali Kosh in the Deh Luran Plain of Iran, where the cereals initially cultivated seem to account for only a small proportion of the food supply and where an increasing reliance on cereal crops can be observed over time. At the site of Tepe Guran, located in a nearby upland area of Iran, it is possible to see a shift at about 6200 B.C. from a semipermanent camp inhabited seasonally by herders of goats to a permanently occupied farming village. At the site of Çatal Hüyük in Turkey, the system of food production has by 6000 B.C. developed sufficiently to support a population of several thousand people.

In general, the following trends in the growth of agricultural systems can be observed: (a) an increasing cultivation of cereal crops; (b) a growing number of species of domesticated animals at sites; (c) a larger percentage of domesticated animals among the faunal remains recovered from sites; and (d) an expansion in the sizes of sites. Among the factors contributing to this economic growth would be the accumulation of human experience related to the handling of domesticated plants and animals and the evolution under artificial selection of the domesticates as biological populations. A less obvious factor would be the interactions in an economic sense between plant cultivation and animal husbandry. For example, there is the possibility of converting or "banking" short-term surpluses of cereals or other crops into animal herds that are temporarily larger in size and that can be used as a buffer against lean seasons or years. At the same time, animal manure can serve to increase the productivity of the land in terms of crop yields.

Another factor contributing to economic growth was the exchange of ideas and probably of seeds and animals between different parts of southwest Asia. Evidence for long-distance exchange networks is provided by obsidian, a volcanic rock that is used for making stone tools. The main sources of obsidian in this part of the world are located in Turkey. By means of techniques such as neutron activation analysis, the source of an obsidian tool can be identified. The argument has been made that items such as obsidian tools moved through networks of sites in southwest Asia primarily by means of exchanges between nearby villages.[13] One implication of such exchange networks is that innovations or successful variations that arose in one place probably had a good chance of reaching other places in the network.

In retrospect, it is evident that the origins of agriculture in southwest Asia involve many different elements. No single variable such as population growth or environmental change can explain the course of this development.[14] Among the elements that deserve special attention are the broadening subsistence strategies of late hunting and gathering groups and the processes involved in plant and animal domestication. Technological developments such as grinding equipment and storage facilities contributed to the successful exploitation of the stands of wild cereals that probably played a central role in the emergence of a sedentary way of life. Sedentism, in turn, appears to have encouraged both population growth,

THE GROWTH OF FOOD PRODUCTION

as we shall see in Chapter 5, and increased pressure on local natural resources. The attempt to manage wild food resources probably set in motion the processes of domestication. The key step in the case of the wild cereals would have been the sowing of previously harvested grain. Agriculture did not suddenly emerge as a fully developed economic system: the plants and animals initially brought under domestication were relatively few in number, and their contribution to the food supply may have been in some cases quite limited. Between 7500 B.C. and 6000 B.C., economic growth due to a combination of factors led both to the progressive modification of the domesticates in biological terms and to their integration in economic systems oriented primarily toward food production.

CHAPTER 3

THE NEOLITHIC TRANSITION IN EUROPE

3.1 European Prehistory on Its Own Terms

A reader turning to European prehistory for the first time must come to grips with a terminology that reflects the growth of archaeology over the last two centuries. The term "neolithic," for example, has several related yet different meanings as we shall see. During the first half of the nineteenth century C. J. Thomsen and J.J.A. Worsaae introduced the classical "three age system"—stone age, bronze age, and iron age—as a means of organizing the collections at the National Museum of Denmark.[1] A characteristic material was used in making the implements belonging to each age. In 1865 John Lubbock in England drew a further distinction between the paleolithic or old stone age and the neolithic or new stone age. This distinction was based on the occurrence of celts or polished stone implements at neolithic sites. Subsequently, a period between the paleolithic and neolithic was recognized for its use of stone tools known as geometric microliths, and the term "mesolithic" was introduced to fill the gap. As chronologies have been further refined, these five ages have been subdivided, and terms such as "early neolithic" and "middle bronze age" are common in the archaeological literature. The basic terms themselves have also been defined in new ways as the result of further investigations. In the late nineteenth century the presence of pottery came to be taken, for example, as a hallmark of the neolithic period in Europe.

In this century, under the influence of V. Gordon Childe and his notion of a neolithic revolution, the term "neolithic" has been redefined and applied to sites displaying subsistence economies based upon agriculture. The term has thus come to possess a combination of economic, technological, and chronological meanings for the archaeologist. In most cases, a correspondence exists among these three dimensions, and the term can be used without ambiguity. But uncertainty about the proper use of the term may arise when one of these dimensions is missing. A case in point would be the sites with pottery but without a farming economy that are found in areas near the Baltic Sea.[2] There is also the opposite case: agriculture is attested at early sites in the Near East, but pottery is absent. In order to deal with this situation, terms such as "prepottery neolithic" and "aceramic neolithic" were introduced. As archaeology has become a more mature science, meanings have evolved and terminology as a whole has become more complex.

This book adopts an economic approach to the classification of sites and cultures: the neolithic transition refers to the shift from hunting and gathering to food production. The term "mesolithic" is used to refer to sites that had subsistence economies based upon hunting, gathering, and fishing. It is worth commenting that the appearance of neolithic economies does not indicate a shift to an exclusive reliance upon farming and stock raising. Hunting and gathering continued to play a part in subsistence economies, as they commonly did among rural populations in Europe down through even historical times. The overall contribution of hunting and gathering, when it comes to meeting subsistence needs at neolithic sites, appears to vary considerably from one region of Europe to the next. This variation seems to be related to decisions made with respect to where sites are located in a region. For example, at sites oriented toward plains that have fertile soils and yet not much ecological diversity in terms of edible wild plants and animals, only modest evidence for hunting and gathering may be documented.

In studying the neolithic transition, it is possible that archaeologists have focused too narrowly on questions of subsistence and have not paid enough attention to other aspects of economic life. Following Childe's lead in redefining "neolithic" in terms of food production, researchers during the last twenty years have dutifully emphasized the study of plant and animal remains recovered from sites and the reconstruction of subsistence economies. More recently, investigators have realized that equally important changes with major economic ramifications took place in nonsubsistence activities. A good example here would be the effort that was often put into the construction and maintenance of buildings at neolithic sites.[3] In economic terms, then, it may be more appropriate to see the neolithic transition as involving broader changes in modes of production, consumption, and exchange. We shall return to this theme in the final section of this chapter.

3.2 THE MESOLITHIC BACKGROUND

Only during the last forty years have mesolithic sites been investigated in any significant number and the character of the period begun to be understood. The boundary between the Pleistocene, or glacial period, and the Holocene, or postglacial period, is conventionally set at 8300 B.C. The cultures dating to the glacial period are commonly referred to as paleolithic, while those dating from the beginning of the Holocene to the appearance of agriculture in a given region of Europe are called mesolithic. In many parts of Europe there is no clear break between the end of the paleolithic and the beginning of the mesolithic: continuity rather than change is observed. Microlithic stone tools that often take geometric forms (see figure 3.1) are the hallmark of mesolithic sites. Although these small flint tools are present in late paleolithic contexts, they become much more common during the mesolithic. Change over time in the typology of the microliths permits the definition of chronological horizons, such as the Sauveterrian and Tardenoisian in France. Several geometric microliths are sometimes mounted together on the same haft in order to produce what are known as composite tools. It is also possible that they served as points on arrows, since there is evidence that the bow was in use during the mesolithic period.

Major cultural adaptations must have been required in the face of the environmental changes occurring at the end of the last ice age. These changes include trends toward warmer temperatures, higher sea levels, and increasing forest cover in Europe. The shift to our present climate actually took place as a series of warmer and colder oscillations during the period between about 10,000 B.C. and 6000 B.C. Not only did temperatures increase during the early postglacial period, but it appears that they may even have reached a level some two degrees centigrade warmer between 6000 B.C. and 3000 B.C. than those prevailing at the present time. The effects of postglacial environmental change were most marked in those regions such as Scandinavia and the Alps where the retreating ice sheets and glaciers opened new areas for the establishment of plants and animals. On the other hand, rising sea levels that were the consequence of the release of water previously captured in glaciers led to the submergence of coastal areas and an overall reduction in the land area of Europe. England and Holland, which had been effectively connected during the late Pleistocene, were, for example, separated as a consequence of rising sea levels and the formation of the North Sea. The higher sea levels had a profound impact on the geomorphology of the lower reaches of rivers and streams feeding into the Atlantic and Mediterranean. Estimates for the rise in worldwide sea levels between 10,000 B.C. and 5000 B.C.

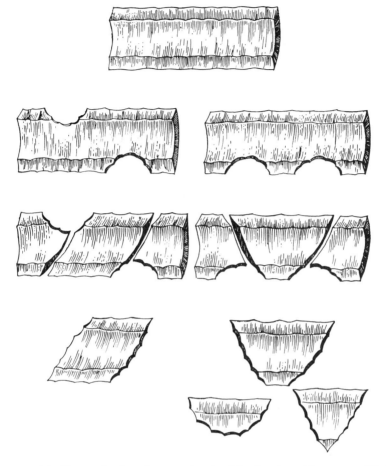

FIGURE 3.1. The production of geometric microliths. Starting with an unmodified blade (at the top), notches are made along the edges and then the blade is broken into fragments. The central fragment on the left can be retouched along its broken edges to produce a rhomboidal form. The central fragment on the right can be flaked to yield a trapeze, lunate, or triangular form (Bordaz 1970: fig. 42)

have been placed at 10 meters to perhaps as much as twenty meters.[4]

From an economic point of view, the most important changes were those related to the effects of climate on vegetation and animal populations. Evidence comes from diagrams of pollen cores dating to the early postglacial period that have been produced for various parts of Europe. These diagrams provide a record of pollen rain in an area, which in turn reflects the changing composition of plant

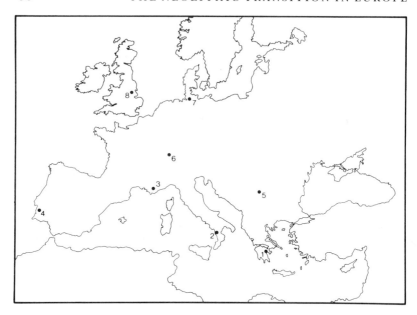

FIGURE 3.2. Selected mesolithic sites in Europe: (1) Franchthi Cave, (2) Grotta della Madonna, (3) Châteauneuf-les-Martiques, (4) Moita do Sabastião, (5) Lepenski Vir, (6) Lautereck, (7) Kiel-Ellerbek, and (8) Star Carr.

communities. What the diagrams document is the eventual establishment of mixed oak forests in many parts of Europe. For animal populations, this change implied a reduction in the space available for large herd mammals with a preference for open habitats, such as reindeer, and an increase in woodland habitats suitable for species such as red deer and wild boar. It comes as no surprise, then, that the latter two species are the ones most commonly represented among the larger mammals at mesolithic sites in Europe.

One of the widely noted features of mesolithic economies is the exploitation of smaller mammals and a wider range of foods, including fish, shellfish, and birds. This shift is usually interpreted as an adaptation to the postglacial environmental changes mentioned above. Increasing forest cover and consequent reductions in herd size made it necessary to turn to alternative sources of food. Again, the shift to a broader resource base appears to have been a gradual one. The subsistence economy at the mesolithic site of Star Carr in England, dating to 8000 B.C., displays a reliance on large herd mammals that is reminiscent of the economies of late paleolithic sites in northwest Europe. By late mesolithic times (4000 B.C.) in this part of Europe, there is a much greater reliance on small game, fish, and shellfish, as seen, for example, at the Ertebølle and Ellerbek sites in the Baltic area. Evidence in Denmark also points to specialized forms of red deer hunting in which only adult animals were selected as prey.[5]

In the Mediterranean, the exploitation of aquatic resources and of shellfish in particular can be tracked back to before 10,000 B.C. Stratigraphic sequences at the Franchthi Cave in Greece and the Grotta della Madonna in southern Italy reveal that fish and shellfish were increasingly exploited during the course of the mesolithic period. The later mesolithic sites of Moita do Sabastião in Portugal and Lepenski Vir along the banks of the Danube show specialization with regard to collecting shellfish and fishing, respectively.[6] It is worth commenting that aquatic habitats are recognized to be among the most productive in terms of their yields of biomass. However, in the context of the total land area of Europe, those places actually well suited to the intensive exploitation of aquatic resources are comparatively limited in number. More recently, it has been suggested by D. Clarke that the collection of plant foods, especially in the Mediterranean area, may have made a much greater contribution to mesolithic diets than has traditionally been thought.[7]

Our knowledge of mesolithic settlement patterns is unfortunately still quite limited.[8] Few sites have yet to produce clear evidence on the plans of dwellings and the techniques used in their construction. Questions concerning whether sites were occupied seasonally or on a more permanent basis remain for the most part unanswered. In general, mesolithic settlements seem to be much smaller in size than their neolithic counterparts. This observation may be a reflection of differences in group sizes, although it may also be related to a greater need for economic space (e.g., the storage of crops and the housing of animals) at neolithic sites. In regions such as central Europe, there is some tendency for mesolithic sites to have different locations than neolithic settlements: in some cases, mesolithic sites appear to display patterns essentially complementary to those of neolithic sites.

3.3 Early Farming Cultures in Europe

The aim of this section is to describe briefly some of the main features of the early farming cultures found in different regions of Europe. For this purpose, it is useful to distinguish five regions: Greece, the Balkans, central Europe, the western Mediterranean, and northwest Europe.[9] Although neolithic subsistence economies

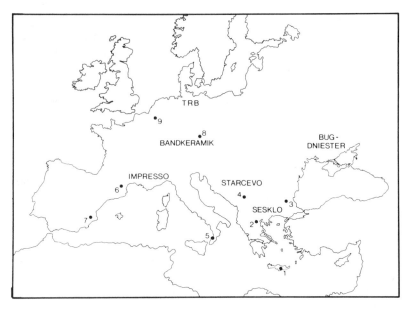

FIGURE 3.3. Selected neolithic sites in Europe: (1) Knossos, (2) Nea Nikomedia, (3) Azmak, (4) Divostin, (5) Piana di Curinga, (6) Abri Jean Cros, (7) Covet de l'Or, (8) Bylany, and (9) Aldenhoven Platte. The names of several of the major early ceramic traditions are also indicated.

display some differences in emphasis from one region to the next, forms of agriculture involving the cultivation of wheat and barley and the herding of sheep, cattle, and pigs are documented in each of these regions. At the same time, it is not uncommon for the early farming sites within a given region to display local variations in their subsistence economies. Such differences are often interpreted in terms of adaptations to local environmental conditions. In broad cultural terms, two aspects of the archaeological record—pottery and settlement patterns—tend to exhibit more marked regional differences. Settlement patterns may differ in the form and size of houses, the size of settlements and the layout of houses within them, and the distribution of sites over the landscape. Often a distinctive vessel shape or style of pottery decoration lends its name to a neolithic culture.

Greece

As well as the modern nation of Greece, this first region includes neighboring areas of southern Bulgaria and the part of Macedonia

TABLE 3.1 Composition of the faunal remains recovered from early neolithic sites in Europe

Site	Date B.C.	Cattle (%)	Sheep/goat (%)	Pig[a] (%)	Dog (%)	Other (%)
Argissa (Greece)	6000	4	83	9	1	3
Knossos (Crete)	6000	16	65	17	2	1
Anza (Yugoslavia)	5250	10	78	8	1	3
Gyálarét (Hungary)	5000	13	29	10	2	46
Jean Cros (France)	4500	11	35	27	1	26
Breść Kujawski (Poland)	4250	54	8	1	1	36

Sources: Argissa (Bökönyi 1974); Knossos (Jarman and Jarman 1968); Anza (Bökönyi in Gimbutas 1976); Gyálarét (Bökönyi 1974); Jean Cros (Poulain in Guilaine 1979a); Bryeść Kujawski (Grygiel and Bogucki 1981).
Notes: The dates are based upon uncalibrated radiocarbon dates (see Site List 4.1) and are cited in terms of the nearest quarter millennium. The percentages are calculated in terms of numbers of specimens.
[a] Both domesticated and wild forms are included here.

located in Yugoslavia. It comprises effectively those areas with rivers that flow into the Aegean. Three names are commonly employed in the literature to refer to the closely related cultures of the region: Sesklo (in Greece), Karanovo (in southern Bulgaria), and Kremikovci (in southern Yugoslavia). Radiocarbon dates (cited here and elsewhere in this chapter as uncalibrated C-14 dates; see figure 4.4) from early farming sites in this region often date back to the sixth millennium B.C. A characteristic feature of the decoration of fine ware ceramic vessels is the use of white painted patterns on a red ground. Such decorative motifs are seldom seen on pottery outside of the region. Settlements taking the form of mounds are another characteristic feature of the region, and the series of stratified habitation levels that is commonly observed at such sites implies a certain continuity in their occupation. In this respect, they show some similarity to neolithic sites in the Near East. The houses in the region are one-room structures, comparatively small in size and generally rectangular. A wooden framework supports walls that are covered with daub. Houses are spaced reasonably close together, and a fair number of houses is usually observed at a given site. The cultivation of a range of cereal crops is attested at these sites, and among faunal remains, domesticated animals predominate, with the bones of sheep and goats the most abundant, followed by cattle and pigs (see Table 3.1). In certain areas of Greece, such

FIGURE 3.4. Examples of early neolithic pottery in Europe: (1) a Criş vessel from Rumania, (2) a cardial impressed vessel from southern France, and (3) a Linearbandkeramik vessel from Germany.

as the plain of Thessaly, dense patterns of neolithic occupation can be observed.

The Balkans

This region corresponds closely with the central part of the Balkans and the system of rivers flowing into the Danube and eventually into the Black Sea. With respect to the names given to early neolithic cultures, Balkanization is again the rule: Starčevo in Yugoslavia, Körös in Hungary, and Criş in Rumania. Radiocarbon dates from early farming sites tend to fall in the late sixth millennium or the fifth millennium B.C. Painted designs based on parallel lines are still seen on fine ware vessels, but the decorative motifs are normally painted in black on a red ground. Anthropomorphic figurines in clay are reasonably common in this region, as they also are in Greece. In contrast, painted pottery and such figurines are not seen at sites belonging to the Bandkeramik culture in central Europe. Settlements in the Balkan region are often located on river terraces where fertile soils are present. Sites can take the form of mounds, as in Greece, or they can occur as comparatively thin deposits, as in the case of many Starčevo sites. Houses again are small, single-room structures having a timber framework covered with daub. Domesticated animals predominate among faunal remains, and, with the exception of sites in Rumania, where cattle seem to have greater importance, sheep and goats are usually the most abundant.

Central Europe

Central Europe is defined effectively by the distribution of sites belonging to the Bandkeramik culture, which is also referred to as the Danubian culture and the Linear Pottery culture. The earliest C-14 dates (uncalibrated) for Bandkeramik sites go back to about 4500 B.C. The striking feature of this culture is its homogeneity

EARLY FARMING CULTURES IN EUROPE   43

over a vast area extending from Hungary to southern Holland and eastern France. The hallmark of the culture is pottery that displays incised "linear" patterns of decoration. As mentioned above, painted pottery is not observed in this part of Europe. The other distinctive feature of the region is its large, rectangular-shaped houses with heavy timber frameworks, which have interior partitions and often measure as much as thirty meters in length. The consistent orientation of houses in a NW–SE direction is indicative of the coherence of the culture over the region. The structures tend to be more widely spaced than is the case in Greece and the Balkans, and it is generally considered that only a few of the houses identified at a site were actually occupied at any one time. There is evidence for the cultivation of cereals, with emmer wheat apparently the preferred crop at many sites. Several archaeologists have argued that a swidden system of farming may have been practiced, which called for the periodic relocation of households.[10] It is of considerable interest that the distribution of Bandkeramik sites is closely associated with the loess soils of the plains of central Europe. These light soils would have been relatively easy to work with a primitive technology. Faunal remains again reveal the predominance of domesticated animals, with a preference for raising cattle documented at most sites. Dense patterns of occupation are observed in areas such as those near Bylany in Czechoslovakia and Aldenhoven in West Germany, where extensive excavations have been conducted. In the latter case, some 160 Bandkeramik houses were identified along one 1.3 kilometer stretch of a stream called the Merzbach.[11] As we shall see in Chapter 5, it is possible to use this evidence to obtain a rough estimate of rates of population growth among early farmers in the area.

Western Mediterranean

The region that includes Italy, southern France, and Spain is the least well-known part of Europe so far as early farming is concerned. One of the main problems is that archaeologists have directed their attention primarily to cave sites, which are often situated in rugged limestone areas that were not well suited for early farming. Only more recently has emphasis begun to shift to the study of open-air settlements. The patterns that emerge from the work at such settlements fit more closely with those seen in other regions of Europe.[12] The pottery of the region is characterized by decorations impressed in the surface of a vessel with the edge of

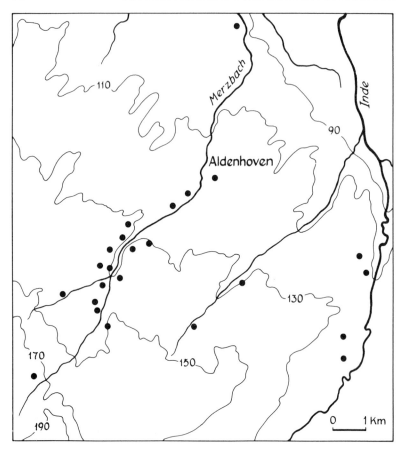

FIGURE 3.5. Distribution map of Bandkeramik sites near Aldenhoven in Germany. It is worth noting the concentration of sites along the stream called the Merzbach.

a shell or some other implement. The term "Impresso" is often applied to the cultures of the region for this reason. Painted pottery is seen only in some parts of southern Italy. Early sites with good evidence for the practice of agriculture have radiocarbon dates that fall mainly in the fifth millennium B.C. On the basis of a few early C-14 dates from sites in southern France, it is possible that pottery was being made in this part of the Mediterranean before farming itself was practiced.[13] There is also some evidence that the herding of sheep and goats may have begun in southern France prior to cereal cultivation.[14] The use of boats in the western Mediterranean is documented by the occupation of a number of islands for the first time during this period and also by the exploitation of island sources of obsidian, a volcanic glass used for making stone tools.[15] Boats would have facilitated movement over substantial distances, as well as contact and exchange between groups living in different parts of the region. Relatively little is yet known about settlement patterns in this region of Europe. Our own recent fieldwork at Piana di Curinga in southern Italy has begun to reveal how houses were built, their number, and their layout at a settlement. Dense patterns of neolithic occupation are observed in various parts of southern Italy, such as the Tavoliere in Apulia and Piana di Curinga in Calabria, that are well suited to the practice of early forms of agriculture.

Northwest Europe

This region includes the north European plain (northern Poland, northern Germany, and Denmark), northwestern France, the British Isles, and Scandinavia. It comprises areas located at the periphery of the spread, where environmental conditions were in general more demanding for the practice of early farming. As one might well expect, a fair degree of diversity is evident among the various cultures in the region. The so-called TRB culture of Denmark and northern Germany takes its name from a characteristic funnel-shaped vessel, *Trichterbecker*, that is decorated with simple impressions. In Britain, early pottery in the Grimston–Lyles Hill tradition has carinated forms, and the vessel surface is finished with shallow fluting. Not only are there differences in pottery styles between the cultures of this region and the Bandkeramik culture, but houses are also much smaller. (It should be mentioned, however, that along the Oder River delta and in a few other places isolated Bandkeramik settlements can sometimes be observed on the north European plain.) One of the distinctive features of the region is the development of a neolithic tradition of monument building that includes burial mounds and dolmens.[16]

3.4 THE NEOLITHIC TRANSITION

The above descriptions are obviously only capsule accounts of the situations in several of the main parts of Europe. A long chapter would be necessary just to summarize the evidence available on each region's early neolithic sites. As mentioned in the preface, our aim in this book is rather to focus on some of the broader themes of the neolithic transition. Although evidence exists for the local domestication of certain plants and animals such as oats, cattle, and pigs in some areas of Europe, local domestication usually took place within economic contexts that included introduced species that were originally domesticated elsewhere. The best examples here would be cereals such as emmer wheat and barley. Little or no evidence has yet been found for the exploitation of wild forms of these plants at mesolithic sites in Europe.[17] As mentioned in Chapter 2, the wild progenitors of the species involved do not appear to have geographic distributions that extend to most parts of Europe. A similar situation appears to hold with respect to sheep and goats in some parts of Europe: the remains of these animals are not commonly recovered from mesolithic sites in such areas.

These patterns imply the introduction of some of the main elements of European agriculture from outside of Europe. It is possible, as we shall see in the next chapter, to map the spread of cultivated cereals in Europe and even to measure the average rate of diffusion of early farming. Before doing so, however, it is useful to make a few general comments about the neolithic transition from the viewpoint of cultural change. A basic question is whether and to what extent continuity with previous mesolithic ways of life can be recognized or whether early farming cultures show a break with preceding local cultures. As the brief descriptions in the last section indicate, major changes in other spheres of material culture seem to have taken place as part of the shift from hunting and gathering to food production. Pottery makes its initial appearance in most parts of Europe in association with the transition.[18] Of perhaps greater significance (and this is a subject to which we shall return below), new patterns of settlement are generally seen with the start of farming in a region.

The class of material culture that offers perhaps the best opportunity for tracing continuity is stone tools, which are incidentally among the most durable and numerous remains recovered from both mesolithic and neolithic sites. In general, there is not much evidence for continuity in stone tool typology or relative abundance of different types within lithic assemblages in Greece, the Balkans, the Bandkeramik area, and much of southern Italy. One observes a primary orientation toward the production of simple blades at early neolithic sites in these areas. The more elaborately worked pieces associated with mesolithic traditions are not commonly seen. In contrast, there seems to be evidence for a greater degree of continuity in lithic traditions at cave sites in northern Italy and southern France. However, it is entirely possible that the lithic assemblages recovered from cave and rock shelter sites may not be representative of the main lines of neolithic development. As mentioned previously, we need to know more about the stone tools from open-air settlements in this area before definite conclusions can be drawn. There is also some indication of continuity in traditions of tool making in certain areas of northern Europe. On balance, the evidence for continuity in terms of lithic remains is at best mixed, being comparatively weaker in eastern Europe than in western Europe.

Perhaps the most sensitive indicator of change is settlement patterns, since they combine aspects of economic and social organization. Neolithic settlements are often placed at new locations on the landscape. Two major factors in site location would have been proximity to water and arable soils. It should be noted that the preferred soils were not necessarily the heavier ones that potentially would have produced high yields but rather in some cases the lighter ones that could be more easily worked with the primitive technologies available.[19] In addition, the amount of arable land required near a settlement in order for its inhabitants to meet their subsistence needs would not have to be very large. It is reasonable to expect that cultivated plots of cereals and other crops were often located within a few hundred meters of a settlement.[20] Unfortunately, the methods of archaeological analysis currently available do not allow us to determine accurately the relative importance of plants and animals as sources of food in neolithic diets.[21] If we knew, for example, the extent of reliance upon various foods and the number of people living at a settlement, rough estimates could be made of the land requirements under early systems of farming. It is worth commenting that the predominance of domesticated animals among the faunal remains at the majority of early farming sites in areas such as Greece and central Europe suggests that meat could easily be obtained without having to resort regularly to hunting.

In terms of the spread of early farming, sites are likely to have been located during the initial phase of settlement of a new area in those places presenting some of the most favorable conditions for the practice of early farming. At such places early neolithic sites would incidentally have had a good chance of success. We might expect, then, that the spatial pattern of the initial settlement of a region would appear as a loosely connected patchwork of occupied areas. Areas offering less favorable conditions might well attract little or no occupation initially and show up as unoccupied islands within the region. Subsequently, occupation would extend to such areas, and over time the region would be gradually filled in with sites.

Another major feature of settlement patterns is the presence of buildings at early farming sites in Europe, which are usually interpreted as houses. With the possible exception of the site of Lepenski Vir on the Danube, where small trapezoidal structures are reported (see note 6 in this chapter), substantial buildings are not observed at mesolithic sites in Europe. The construction of houses with timber frameworks and earth-covered walls would have required a certain investment of time and labor. In addition, a good deal of effort may have been required to procure building materials. The houses, especially the larger ones seen at Bandkeramik sites in central Europe, also suggest a development with regard to the social organization of labor, since individuals outside of the immediate family would have been required for certain tasks. But perhaps most important, the houses imply a more sedentary way of life: this innovation, more than agriculture itself, may represent the main transformation in life style associated with the neolithic transition. Sedentism sets the context for the intensification of a wide range of economic activities—farming, house building, the mining of flint for stone tools—and at the same time permits the elaboration of social organization. It is worth recalling here that the much larger sizes of neolithic settlements in comparison with mesolithic sites would seem to indicate the emergence of larger social units. The argument has been made that the economic change occurring at early farming sites in Europe is linked with the conduct of new forms of social life.[22] We may well have to look beyond subsistence in order to comprehend the wider organizing principles of neolithic economics.

Once they had made their appearance, the early farming cultures in certain areas of Europe seem to have persisted for substantial periods of time without experiencing major changes. In some cases there is even evidence for more or less continuous occupation at the same early farming settlement for several hundred years. This continuity suggests a certain stability so far as cultural and economic adaptions are concerned, at least among certain early farming populations in Europe. Although early farming cultures undoubtedly experienced internal growth and development, the pace of cultural change in a given locality often appears to have been quite slow, at least when compared with subsequent prehistoric cultures in Europe, which seem to manifest more active tempos of change.

Not enough attention has been paid to the question of why stability and persistence in cultural forms, sometimes lasting for a period of several hundred years or more in certain areas of Europe, should ensue from a transition that involved the introduction of essentially new cultures and subsistence economies. The early farming cultures in Europe also appear often to have covered extensive areas in geographic terms, especially the Bandkeramik culture in central Europe. If we turn to more recent prehistoric periods, we see much smaller cultural areas on the map of Europe. Interesting questions can be raised about the overall pattern of differentiation of early farming cultures within Europe. Although ecology definitely seems to have played a leading role, other factors—many of them still poorly understood at the present time—undoubtedly contributed to the spatial and temporal coherence often observed among early neolithic cultures in Europe.

CHAPTER 4

MEASURING THE RATE OF SPREAD

Only in the last fifteen years have archaeologists begun to take a more quantitative approach to the reconstruction of the past. When we began working in collaboration in 1970, it was still quite unusual to apply quantitative methods to a research question such as the spread of early farming in Europe. Initially, we turned to the problem of measuring the actual rate of spread by using radiocarbon dates from early neolithic sites in Europe. A rate measurement, even as a rough first approximation, might provide some new insight into the processes involved in the expansion. Somewhat to our surprise, we found that a rate measurement could be obtained: there is a reasonably consistent pattern to C-14 dates from early neolithic sites when they are plotted on the map of Europe. The results of this initial study were presented in a 1971 article entitled "Measuring the rate of spread of early farming in Europe." We then turned to a more refined approach for displaying the pattern, which involved using a computer to generate isochron maps and which yielded essentially the same results.

What made the measurement of a rate possible was the radiocarbon method of dating prehistoric sites developed by W. F. Libby at the University of Chicago just after the Second World War. Previously, it had been possible to date neolithic sites in Europe only in terms of relative chronologies. By the mid-1960s, a fair number of C-14 dates had become available for neolithic sites, and for the first time one could get a rough idea of the absolute ages of the early farming cultures in different parts of Europe. Of the various early attempts at collecting this data, perhaps the best known is that published by J.G.D. Clark in 1965 (see figure 4.1). Here symbols are used to represent three different time intervals, and the overall pattern reveals an east-to-west trend. Clark's map thus confirmed the widely held view that agriculture began in the Near East and spread first to Greece and the Balkans and subsequently to central and northern Europe. But no serious attempt was made to use the information available in a more analytical way. The problem was in large part one of developing a method of making a rate measurement.

THE INITIAL ANALYSIS

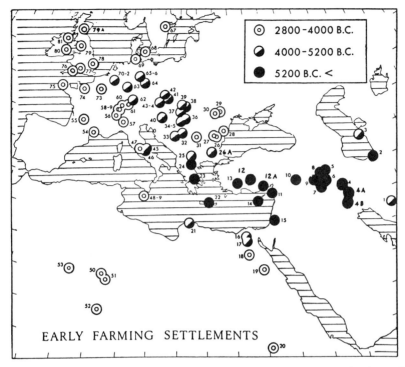

FIGURE 4.1. Map of early farming sites published by Clark in 1965. The dates of the symbols are in uncalibrated radiocarbon ages.

4.1 The Initial Analysis

It is worth reviewing the steps in our thinking as we began to work on the question of measurement. One of the first steps was to clarify what it is that one should attempt to measure. Evidence for early farming is usually found at archaeological sites that are called neolithic. As we saw in the last chapter, a whole complex of traits can be observed at neolithic sites, and yet there is some variation in the elements of this complex from one region of Europe to the next. Although it appears that what is spreading in most parts of Europe is the complex as a whole, it is undesirable to measure the expansion of early neolithic sites as such, owing to the different ways in which the term "neolithic" is defined in various parts of Europe. Instead, it would be preferable to base the rate measurement on a given trait that is of central importance to early farming. Cereal crops are in many ways more useful in this respect than domesticated animals, since they offer a much clearer morphological basis for distinguishing domesticated from wild forms.

As we have seen in Chapter 2, the natural geographic distributions of the wild progenitors of cereals such as emmer wheat and barley do not extend to most parts of Europe. With the possible exception of the Franchthi Cave in Greece, no serious claims have been advanced for the local domestication of wheat and barley in other parts of Europe. On the other hand, evidence exists for the occurrence of domesticated forms of wheat and barley at early neolithic sites in various parts of Europe. If possible, one would like to obtain C-14 dates from charred cereal remains at early farming sites in Europe. This ideal approach has indeed become feasible during the last few years with the development of new methods for making radiocarbon determinations on very small samples. Prior to 1978, the sample sizes required for C-14 determinations made it impractical in most cases to date cereal remains directly, since they usually are recovered in small quantities and often form a valuable part of the paleobotanical record. Thus most dates were determined from fragments of charcoal recovered from sites. Positive evidence for wheat and barley has been found at many neolithic sites in Europe, but not at all sites, particularly those that were excavated prior to the 1960s, before the introduction of flotation methods for the recovery of carbonized plant remains. In order to obtain broad geographic coverage in Europe, we decided to include certain neolithic sites in the analysis even though they did not have positive paleobotanical evidence for the cereals. We based this decision on the consideration that when conditions of preservation are favorable and appropriate techniques of recovery are employed, cereal remains are almost always found at neolithic sites in Europe. In other words, it was desirable to work with as large a sample of dated sites as possible.

The real question then became how to carry out the analysis. A pioneer attempt by M. S. Edmonson to measure neolithic diffusion rates with regard to the appearance of pottery in various parts of the world suffered from an approach that was unsatisfactory in many respects.[1] The problem would be simpler if diffusion were taking place in a space with essentially one dimension, as in the case of a coastline or river. In practice, however, the pattern of diffusion is usually more complex. Even though paths of movement may be largely conditioned by certain geographical factors, there is usually more than one route along which diffusion from one point to another can take place, especially if the two points happen

THE INITIAL ANALYSIS

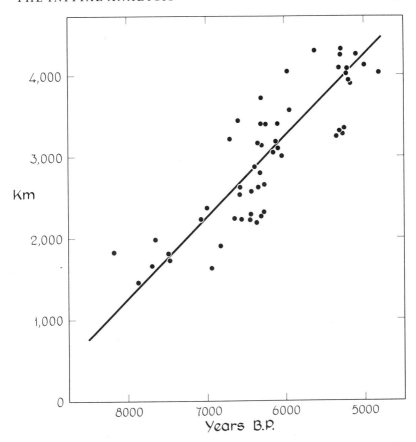

FIGURE 4.2. Regression analysis of the spread of early farming in Europe. The site of Jericho is taken as the presumed center of diffusion in this case. The points correspond to sites providing estimates of the time of arrival of early farming in different parts of Europe. The distance of a site from Jericho (on the ordinate) is measured as a great circle route. The dates of sites (on the abscissa) are conventional radiocarbon ages in years B.P. (after Ammerman and Cavalli-Sforza 1971: fig. 2).

to be located some distance apart. This difficulty can be overcome if a center from which the process originates can be defined; then distances are measured from the center to various points on the map, providing estimates of the time of arrival of the element that is spreading. One can then construct a graph such as the one shown in figure 4.2, where arrival times at given points are plotted against their respective distances from the center.

Such graphs may take several different forms, and it is useful to consider the pattern that a plot can be expected to have in the simplest case, as well as the effects that various complications may

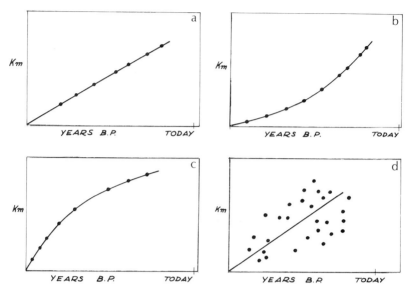

FIGURE 4.3. Expected patterns of diffusion under four hypotheses. The plots show distance from a center of origin against time in the case where: (a) the rate of diffusion is constant in time and space, (b) it increases with time, (c) it decreases with time, and (d) it is variable in space.

have on the pattern. In the simplest case (1), the rate of diffusion would be constant in both time and space, and the expected pattern would be a linear relationship (as shown by *a* in figure 4.3). The agreement between this hypothesis and the data can be evaluated by means of the linear correlation coefficient r, which in the case of perfect agreement will have a value of 1.00. Here the rate of diffusion is measured by the slope of the regression line that has been fitted to the points of the graph. In an alternative case (2), the rate of diffusion may not be constant but rather may increase or decrease over time. Curved lines such as those shown in *b* (increase) and *c* (decrease) of figure 4.3 would be expected where the diffusion rate takes on different values at different points in time. In another case (3), the rate of diffusion may vary in space, producing a scattered array of points. The degree of scatter will be a function of the variation in the rate. An example of a reasonably large variation of this kind is represented by *d* in figure 4.3. Deviations from case (1) due to the complications of either case (2) or case (3) will be reflected in lower values of the correlation coefficient. A further case (4) should also be considered: namely, one in which the actual location of the center of diffusion is not well known. In theory, if the data are abundant, it would be possible to compute correlation coefficients for a large number of different potential centers, with the one yielding the highest value of the coefficient being taken as the most likely candidate for the center.

Case (5) would occur where a single center is used for plotting the graph when there are in fact several independent centers of diffusion. The expected pattern, if the plot is made with all of the distances measured from only one of the centers, will show a wide variation in space and time. In general, values of the coefficient should be low and could even be negative, indicating that the hypothesis of a single center is wrong. For purposes of computation, the center must be regarded as a single, isolated point. It is much more realistic, however, to consider the case (6) where the center is expected to cover some area rather than being simply a point. The distortion in the graphic representation that may arise in this case will be of importance only for those points located near the center and will become negligible as one moves away from it. In the example of an actual plot shown in figure 4.2, all of the measurements of distance happen to be of the latter kind; that is, the sites plotted are not in immediate proximity to the central area. The pattern in this case can be expected to be similar to that of case (1) if the rate does not vary in time or space.

We have avoided using starting dates for the diffusion process, since there is relatively little chance that we will have good direct evidence for assigning them. It is worth stressing here that the diffusion process should not be confused with the separate but related one of domestication. Diffusion will probably begin from the same general area where domestication occurs, but we are unlikely to know the exact temporal relationship between domestication—which is itself difficult to pinpoint in time—and the effective starting date of the diffusion process. In contrast, it is much easier from a sampling point of view to recognize points in Europe that provide estimates of time of arrival, since there are usually marked differences between early neolithic sites and mesolithic sites in a given area.

Our original analysis was based on a group of 53 sites with radiocarbon dates. Whenever more than one layer or level of a site was dated, the earliest neolithic one was used in the analysis. Sites were included in the analysis only when their radiocarbon dates had standard errors of two hundred years or less. Excluded from the analysis were sites located in the Alps, where retardation of the spread was expected owing to unfavorable ecological and geo-

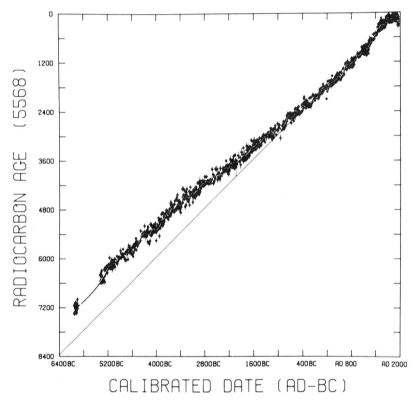

FIGURE 4.4. Radiocarbon dates and calendric dates over the last 8,000 years. The figure is based upon the analysis of samples of bristlecone pine and giant sequoia. A sample's conventional radiocarbon age in years B.P. is given on the ordinate. Its associated calendric date, as determined by dendrochronology, is plotted on the abscissa. If conventional C-14 years were equivalent to calendric years, all of the data would fall on the diagonal line (Klein et al. 1982: fig. 2).

graphical factors. In the graphs prepared, the dates are radiocarbon ages in years B.P., (before present), as they are conventionally reported in the journal *Radiocarbon*. No attempt was made to correct the C-14 dates using calibration curves, which were still at an early stage of development and did not extend back much before 6000 B.P. Main trends in the relationship between C-14 dates and calendric dates over the last 8,000 years are shown in figure 4.4. Analysis of dendrochronologically dated samples of bristlecone pine and giant sequoia provides the data for constructing such curves.[2] Over the time period that interests us and that is covered by currently available calibration tables (i.e., 7200 B.P. to 4800 B.P.), it can be seen that C-14 dates at 7200 B.P. tend to lag on average about

THE INITIAL ANALYSIS

750 years behind calendric dates, while those at 4800 B.P. are some 850 years behind calendric dates. That the lag at the two ends of this time interval (lasting 2,400 radiocarbon years) is approximately the same means that the eventual calibration of dates is not likely to have an appreciable effect on the slope of the regression line. The calibration of dates is likely to have a greater effect on the values of correlation coefficients, however. It should be noted that since calibration tables go back only to about 7,200 B.P. at present, they do not cover the full range of C-14 dates obtained for sites in Greece and the Balkans (see Site List 4.1).

Four sites in southwest Asia—Jericho, Çayönü, Jarmo, and Ali Kosh—as well as a location representing the center of gravity in geographic terms of these four sites, were chosen for making graphs of the kind shown in figure 4.2. Distances to sites in Europe were computed as great circle routes using latitude and longitude coordinates. High values of the correlation coefficient were obtained in all five cases, ranging from $r = 0.89$ for Jericho to $r = 0.83$ for Jarmo and Çayönü. The results of the analysis indicate that the rate of spread was reasonably constant over a wide range of time and space. An estimate of the overall rate of diffusion is supplied by the slope of the straight line. For the principal axis, which seems to be the preferred regression line in this case, the slope gives a rate of about one kilometer per year.[3] The linear relationship of figure 4.2 also permits the extrapolation of a starting date where the line crosses the abscissa at zero distance. In the case of Jericho, the date turns out to be approximately 9500 B.P., or 7500 B.C., which would seem to be entirely reasonable. This indirect estimate of a starting date serves as a check on internal consistency. It is worth stressing that the rate of spread that is being measured is an average one for Europe as a whole. We do not claim that the spread was occurring at a single, uniform rate at all points in Europe. Rather, as in the case of a car in a grand prix race, which will exceed and fall below its average velocity at different points in time, the rate of spread would be expected to experience some local variations and the average rate to provide a central value for the spread as a whole.

There is also the possibility that the rate may exhibit some degree of variation from one region to the next. This question was examined by making separate rate estimates for different parts of Europe. The data available at the time were quite limited for computing such regional rates, and the estimates obtained are to be viewed largely in heuristic terms. We found that the rate for the

Balkans seemed to be somewhat slower than the average rate, while that for the western Mediterranean was somewhat faster. The rate may have been even faster in those parts of central and western Europe occupied by the Bandkeramik culture, although the high value here may be due in part to a lack of early dates for the eastern part of the Bandkeramik distribution.

4.2 FURTHER ANALYSIS

Starting in 1973, we began to explore other, perhaps more refined ways of studying the pattern and rate of spread. One approach that appeared to be promising entailed using the dates of sites to draw a series of isochrons, or lines representing the same times, on a map of Europe.[4] This method is analogous to drawing contour lines through points with the same elevation on topographic maps. In our case, the computer-generated lines are based on a fitting procedure that operates directly on a geographic array of dated sites. Such an approach allowed us to avoid attributing centers of origin for the spread. It was no longer necessary, as in the original regression analysis, to measure the distance between a site in Europe and a presumed center in the Near East. The approach also appeared to afford a better reflection of regional variations in the rate of spread, which would be seen in terms of the spacing between the isochrons on the map: isochrons run closer together when the rate of advance is slow and further apart when it is faster. In examining the isochron maps, it would be of particular interest to look at the pattern of spacing between successive lines.

No less important was the possibility of using a greater number of sites in the analysis, since new C-14 dates had become available. The isochron map shown in figure 4.5 is based on a total of 106 sites (see Appendix, Site List 4.1). The lines drawn here at 500-year intervals run parallel to one another over much of Europe, implying a basic regularity in the rate of expansion over a broad range of space and time. The estimate for the average rate of spread between Greece and Britain again turns out to be close to one kilometer per year. As expected, regional variations such as the delay in the Alpine area, the faster expansion in the area occupied by the Bandkeramik culture, and the comparatively slower spread into areas such as Denmark are clearly revealed. It is worth commenting that coverage is still thin in terms of dated sites for some parts of the map, notably the Iberian Peninsula, Rumania, and the

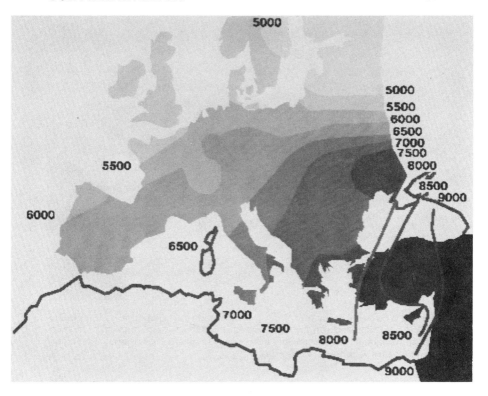

FIGURE 4.5. Isochron map of the spread of early farming in Europe. The isochron lines are drawn at 500-year time intervals. The dates are conventional radiocarbon ages in years B.P. In making this computer-generated map, 106 sites were used (see Site List 4.1).

Ukraine. The map as a whole should be viewed only as a current approximation. As C-14 dates become available for more sites, and as our knowledge of local neolithic sequences increases, we can expect progressive improvements in this approximation.

We can also look at the neolithic transition in quite a different way: that is, from the perspective of the survival of hunter-gatherer or mesolithic populations in Europe. A question of obvious interest is what pattern emerges when the dates of the "latest" mesolithic sites in various parts of Europe are mapped. Such a map draws essentially on an independent data set and therefore might offer an indirect check on the rate measurement. The isochron map shown in figure 4.6 is based on a total of 62 sites with C-14 dates

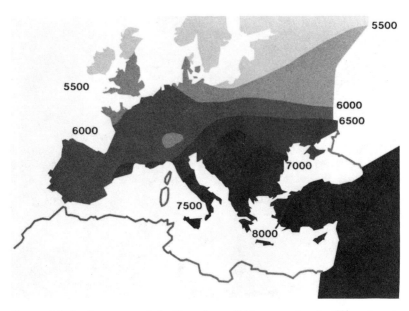

FIGURE 4.6. Isochron map of the "latest" mesolithic occupation in different parts of Europe. The isochron lines are drawn at 1,000-year time intervals. The dates are conventional radiocarbon ages in years B.P. In making this computer generated map, 62 sites were used (see Site List 4.2).

(see Appendix, Site List 4.2) providing estimates of the latest mesolithic occupation in different areas of Europe. A comparison of the two isochron maps shows a good *overall* correspondence between them, in part perhaps because most areas of Europe witnessed no prolonged chronological overlap between mesolithic occupation and the onset of early farming. One implication is that complex population interactions may be involved. (See section 7.3 for a further general discussion of this topic.)

4.3 INTERPRETING THE RATE OF SPREAD

To the modern reader, the most striking feature of the overall spread may well be the slowness of the rate. Even when we think about events taking place in the historical past, we are accustomed to cultural changes occurring at a much faster pace. Early farming took more than two thousand years—the full length of the Christian era—to travel from Greece to the British Isles. By the time cereal crops were first cultivated in northern Europe, they had been sown and harvested well over a thousand times in many areas of southeastern Europe. One implication of the slowness of the rate is that there was ample time for adaptation to local conditions during the course of the spread. There was also considerable opportunity for evolution to occur within neolithic cultures and for various regional cultures to develop, as discussed in the preceding chapter. On the other hand, a rate of one kilometer per year means that the frontier zone where agriculture is being practiced for the first time advances on average some 25 kilometers every generation. In the context of the technology and means of transportation in use among early farming communities in Europe, such a rate is not all that modest. The tempo of events in those places where the spread was actually taking place at a given point in time was probably more active than the rate estimate might lead us to think.

As mentioned before, we can view the spread of early farming as a process of diffusion that can be explained in two essentially different ways. A clear distinction needs to be made between them at the conceptual level. According to one, the process may be seen as the result of *cultural* diffusion. Cereals and farming methods would be passed from one group to another without involving the movement or geographic displacement of people. Alternately, the spread can be seen as a diffusionary process brought about by population growth and displacement. The latter explanation can be referred to as *demic* diffusion. In this case it is not the idea of farming that spreads but the farmers themselves—and with them, their culture. The demic mode of diffusion will be most relevant in situations where a marked change in population level is a consequence of the new form of subsistence. As we shall see in the next chapter, there seems to be a clear association between the neolithic transition and population growth. Moreover, it can be shown mathematically that if an increase in population coincides with a modest local migratory activity, a wave of population expansion will ensue and progress at a constant radial rate. Such a form of expansion is precisely what we observed in the case of the rate measurement of the European data. A more detailed discussion of a model that predicts such a "wave of advance" will be developed in the next chapter. This type of demic diffusion may be distinguished from "colonization," which in its conventional meaning entails the intentional settlement by a group of people of some size, usually in a distant land. A familiar example of colonization is that of the western Mediterranean, recorded in classical

Greek history. By contrast, the model of a population wave of advance would be one of slow, continuous expansion, involving the frequent formation of new settlements at short distances from previous places of occupation.

The two modes of explanation lead to quite different expectations with regard to cultural patterns. In the case of the cultural mode, the actual process of diffusion might take a wide range of forms, but one would expect a certain continuity in material culture and settlement patterns between late hunter-gatherer (mesolithic) sites and early farming (neolithic) sites in a given area. One might also expect a fair amount of cultural variability between one region and the next. In contrast, there would be less expectation of continuity between late mesolithic and early neolithic ways of life under the wave of advance model. We might also expect early farming cultures to cover reasonably large areas and to reflect a lesser degree of variability at the regional level. The latter two expectations seem to hold, for example, with reference to the Bandkeramik culture.

It is not our aim in this section to evaluate various regions or cultures in terms of these general expectations. In the demic case, we also need to consider the interactions that can occur between a population of farmers expanding into a new area and the population of hunters and gatherers already living in the area. Such interactions might take any one of many different forms, ranging from mutualism through acculturation of the local mesolithic population. Interaction is likely to vary to some extent from one region to the next, and detailed archaeological evidence will probably be needed before much progress can be made on this question.[5]

The two modes of interpretation—cultural and demic—that we have briefly discussed need not be regarded as mutually exclusive. In fact, in trying to account for the spread of early farming, one of the more productive research strategies is to try to evaluate the relative importance of the two within various regional settings. But what is necessary before such an attempt can be made is the introduction of much more specific models. At this point we need to discuss in greater detail the wave of advance model.

CHAPTER 5

THE WAVE OF ADVANCE MODEL

5.1 The Neolithic Rise in Population

In trying to explain the observed spread of early farming in Europe, it is interesting to note that substantial changes in population densities often took place concurrently with the shift to agriculture. This suggests a possible connection between growth and spread. It is widely recognized that the shift from food collection to production allows a substantial increase in the number of people that can be supported in a given region or, using an ecological expression, in the region's carrying capacity.

Turning to the ethnographic record, we find that, in spite of considerable variation, densities of hunter-gatherers are generally much lower than those of agriculturalists. With the exception of coastal fishers of the Pacific Northwest, the former have densities that range for the most part between .01 and 1 inhabitants per square kilometer. Densities for agriculturalists range from about 3 to 288 inhabitants per square kilometer.[1]

It is commonly held that the neolithic transition has, over the long run, produced a 100-fold increase of population densities. In order to trace the trajectory of population growth in detail, we must turn to the archaeological record. Although good qualitative evidence exists for population increase associated with the neolithic, as seen in terms of the greater number of sites and also the larger sizes of settlements, there are only a few areas in Europe where density levels and growth rates can actually be measured. One of these, the Aldenhoven Platte in Germany, which we shall discuss in section 5.3, suggests a local density of more than two people per square kilometer and an initial growth rate on the order of 1.5% per year. Archaeologists have only recently become actively interested in developing data sets and methods of analysis for describing in quantitative terms the major demographic shift associated with the neolithic transition.

The occurrence of population growth during neolithic times raises interesting questions about how and why populations actually grew. Ultimately, growth of a closed population can occur only when birth rates exceed death rates. Leaving aside migration, we know that the rate of population growth is computed on the basis of the number of births minus the number of deaths; thus an increase in population size must be due to increasing birth rates, decreasing death rates, or some combination of the two. In order to gain an idea of which change accounted for growth in neolithic times, it is useful to turn to demographic evidence from the archaeological record. Skeletons from burials can be aged and sexed (even if there are limits on precision), and information on skeletal populations can be used to construct distributions of age at death and even life tables.[2]

Considerable variation in the expectation of life emerges from different archaeological series. This result is not too surprising, since the samples involved are often comparatively small and several assumptions need to be made in building such life tables. By and large, however, no major change in the expectation of life at age 15 seems to have accompanied the transition to agriculture (see figure 5.1). Until the last three centuries, it would appear that no major change occurred in mortality rates of human populations. One might expect a decrease in mortality to be associated with the transition, if agriculture ensured a more reliable food supply. But, as mentioned earlier, the nutrition of hunters and gatherers is in most cases qualitatively and quantitatively quite satisfactory.

If there was no major change in death rates, then increasing birth rates must be the main source of growth. Ethnographic evidence shows that the fertility of hunter-gatherers is low in comparison with that of farmers. Typically, hunters and gatherers have a spacing of four years on average between successive births and a completed fertility (i.e., for women surviving to menopause) of five children. With the mortality rates that are prevalent among hunter-gatherers, births and deaths tend to balance one another so that such populations are basically stationary from a demographic point of view.[3] Modern hunters and gatherers probably represent the latest heirs to a reproductive strategy that allowed human populations to grow only at a very slow pace and rarely led to overcrowding. R. B. Lee has suggested that a four-year birth interval would have been highly adaptive in the context of the nomadic life style of hunter-gatherers.[4] In such a society, it is convenient for a woman to carry no more than one small child whose age is such that he or she cannot walk at the parents' pace; a four-year birth interval would ensure that no other child is born before the last born is sufficiently independent. The shift to sedentism with agriculture removes this constraint and makes it possible to

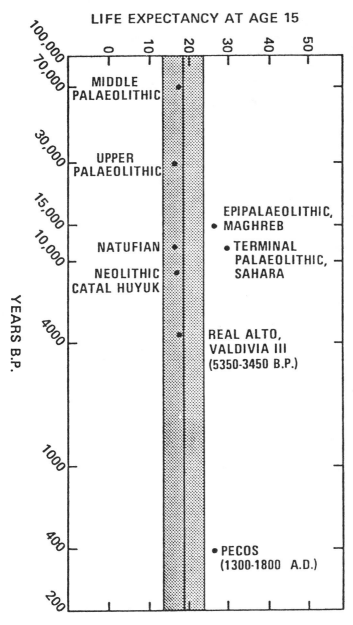

FIGURE 5.1. Life expectancy at 15 years of age among populations living at different periods of time. The estimates of life expectancy are based upon the analysis of sets of skeletal remains ranging in age from middle paleolithic to modern times (after Hassan 1981).

shorten the spacing between births and thus have a larger number of offspring. In the ethnographic record it is common for contemporary populations that are primarily engaged in farming to have a birth interval on the order of 2.5 years. Another factor that may have increased the incentive for a greater number of children would be the more useful role that they can play in agricultural economies. This advantage and sedentism in combination may well account for the shift to higher levels of fertility during neolithic times.

It is worth contrasting the neolithic rise of population with the so-called demographic transition that has occurred in Europe and North America over the last 200 years. The latter is characterized by a regular and substantial decrease in both death rates and birth rates, down to the contemporary low levels in developed countries. The change in death rates is believed to be a consequence of improved sanitation and standards of living (e.g., more and better food and improved water supply).[5] The decrease in birth rates is less well understood. Prior to this demographic transition, which is now slowly extending to the rest of the world, all Western countries were—and many other countries still are—reproducing with high fertility rates. Among rural European populations at the beginning of the nineteenth century, the average birth interval may have been close to 2.5 years.[6] Thus the neolithic rise of population and the modern demographic transition can be seen as two major episodes in human population history that occurred in essentially different ways. It would appear that increased birth rates are primarily responsible for population growth in the first case. In the second case, birth and death rates both decrease, and the growth rate eventually tends toward zero. But in countries where the decline in the birth rate lags behind that of the death rate there is a short but intense period of population growth.

Specific models of population growth will be discussed in section 5.3, but it will be intuitively clear that growth over the long term has its limits. When a population reaches an upper limit, it may remain at that "saturation" level. In the absence of room for expansion the growth rate will ultimately fall to zero. Alternatively, growth may take place under conditions where expansion to adjacent areas can occur at the same time. Conceptually, it is useful to distinguish three cases as a means of better understanding the interrelationships between processes of population growth and expansion. The first would be the one mentioned above, in which there is sustained growth but expansion is not allowed to occur, leading to an increasing buildup in local population densities. In the second case, the population is not growing but does experience expansion; local population densities will consequently become thinner and thinner. The third case is the one in which both growth and expansion occur. Local population densities will tend to increase, and at the same time the total area occupied will also become larger. The first two cases are essentially unrealistic. A population is not likely to concentrate entirely in one area if adjacent space and resources are available. Nor can we expect a population of constant size to experience a sustained expansion, since thinning is likely to lead to extinction. The wave of advance model, to be described in the next section, provides a quantitative framework for considering processes of growth and expansion simultaneously.

5.2 The Model

In the wave of advance model, population growth and local migratory activity are seen as producing a diffusionary process that takes the form of a population wave expanding outward at a steady radial rate. As we shall explain, the model also specifies the quantitative relationships that are considered to hold among three variables: that is, the rate of population growth, the rate of migratory activity, and the velocity of the spread. We first introduced the model in a 1973 paper, "A population model for the diffusion of early farming in Europe." It represents one of the first attempts at formal model building in archaeology. In fields such as population biology and genetics, the use of mathematical models has a longer tradition. As conceptual constructs, formal models provide us with a means of describing complex events in a concise and structured way. They take as their premise a need to simplify complex events or situations in order to place them within a predictive framework. This framework in turn offers a means for evaluating how well a given model fits observations drawn from events that have actually occurred. In proposing a quantitative model, one is forced to expose the weaknesses of a system of hypotheses and to narrow down the set of alternative formulations to those that are compatible with the ensemble of available observations.

It can be seen in retrospect that there was a major gap in the repertoire of models available to archaeologists for explaining cases of diffusion involving population movements. Terms such as "migration" and "colonization," which we can now see were invoked all too often in the earlier prehistoric literature, tend to have connotations of movement over some distance, a planned expedition, and a basic change in cultural context. No model adequately described situations where movements were over short distances for the most part (e.g., the relocation of settlements to adjacent, unoccupied land) and where the cumulative result of many such movements was a slow and continuous form of expansion. In such a case, it would be possible for the "migrants" to maintain a continuity in their social contacts and also in their cultural context. The wave of advance model makes its contribution by suggesting how local processes such as population growth can produce what in some respects is a form of colonization without colonists.

The wave of advance model has been in existence since 1936 and was first put forward by the British geneticist R. A. Fisher to predict the spread of an advantageous *gene* in a linear habitat. The ecologist J. G. Skellam used the model in 1951 to predict the spread of a *population* in two dimensions. The mathematician D. G. Kendall independently developed similar models in 1948 for the propagation of epidemics. The differential equation proposed by Fisher for representing the propagation of an advantageous gene can be applied without change to the geographic spread of populations.[7] In order to predict the number of individuals living at a given point in time and space, one must adopt specific treatments of both growth and migration. In particular, Fisher's model makes two main assumptions. First, growth occurs in a "logistic" manner: initially, the growth of a population takes place actively, but gradually it slows down over time as the population increases in size; eventually, growth stops when the population reaches a "saturation" level. Second, migratory activity takes place at a constant rate in time and according to a "random walk" process. The specific meaning of these assumptions will be explained later in this chapter.

On the basis of these two assumptions, Fisher predicted that a "front" or "population wave" would soon form, and it would keep advancing at a rate of r not less than $2\sqrt{ma}$, m being a measure of the migratory rate and a that of the initial growth rate. Only recently have mathematicians found a solution to Fisher's equation, showing that the constant in front of the term $\sqrt{ma}$ is in fact $5/\sqrt{6}$, or 2.04.[8] This solution refers specifically to a one-dimensional habitat. Although it is still unclear what mathematical form the extension to two dimensions will take, a simulation study described in Chapter 7 shows that a regular front does form and advances effectively at a constant rate. The rate of advance turns out to be not far from that valid for the one-dimensional case. In the present uncertainty,

THE MODEL

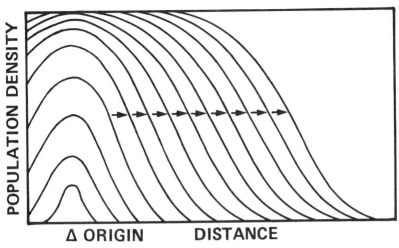

FIGURE 5.2. Fisher's model of a population wave of advance. This graphic representation shows the rise in local population density expected with increasing distance from the origin as time elapses.

we have calculated the curves shown in figure 5.8 on the basis of the recent analytical solution for a linear habitat mentioned above.

The two assumptions of the basic model—logistic growth and random walk migratory activity—can be modified without necessarily compromising the main result: namely, a wave front that advances at a constant radial rate. However, the actual value for the rate of advance is likely to change in such cases. One can imagine many possible variations in the treatments of the two components. Some of these, such as a tendency toward centrifugal movement rather than random walk migration and the presence of barriers in the plane of movement, have been considered in simulation studies and shown to produce only second-order effects.[9]

Among the few applications to date of Fisher's model, one study is worth mentioning: that by Skellam of the spread of muskrats in central Europe. This species, whose fur is of industrial importance, occurs naturally only in North America. Attempts were made at the end of the last century to breed muskrats in captivity, and animals were exported to Europe for this purpose. Once there, a few muskrats escaped, found the plains of Europe to their liking, multiplied freely, and established themselves as a feral species. Through growth and dispersal, they rapidly covered a wider and wider area. The appearance of the species in different parts of Europe can be mapped at various points in time, as shown in figure

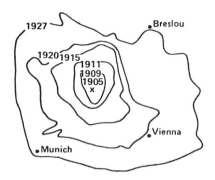

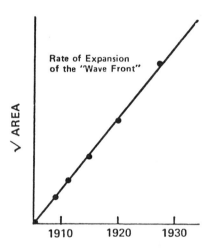

FIGURE 5.3. The spread of the muskrat (*Ondatra zibethica* L.) in central Europe. *Above*: Areas in which the muskrats were found in central Europe. *Below*: The square root of the area as plotted against the year (data based on Ulbrich 1930). The figure is taken from Skellam 1951.

5.3. The contours of this map are highly irregular, as might be expected in the case of an animal species spreading through a highly heterogenous environment. But if the law of averages is given a chance to operate, regularities emerge. Skellam measured the area covered at a given time and plotted its square root against the year, obtaining a surprisingly good fit to a straight line. It is worth recalling that the square root of the area of a circle is proportional to its radius or the average distance from the origin to the expanding front. In other words, the area increases propor-

TABLE 5.1 Rates of population growth and the time required for a population to double in size

Growth rate per year	Doubling time in years
3.00%	23
2.00%	35
1.00%	69
0.30%	231
0.10%	693
0.03%	2,310
0.01%	6,931

Note: The model of growth considered to operate here is an exponential one.

tionately with time, as Fisher predicted, and the slope of the line measures the rate of advance of the wave front. The rate can be estimated at about twenty kilometers per year, or some twenty times faster than the spread of early farming in Europe. If we knew the rates of growth and migratory activity for the muskrat, we could evaluate how well Fisher's model predicts the observed rate of expansion of the muskrat. Not much is known about these two aspects of muskrat populations, however. We are in a somewhat better position, though, when it comes to our knowledge of rates of growth and migratory activity among human populations.

5.3 Logistic Growth

The simplest form of growth occurs when a population increases at a constant rate over time. This is often called exponential growth and will lead to a population enlarging geometrically in size over time. In the case of human populations, we know that the fastest rate of growth is about 3% per year, which means a doubling time of about twenty three years. But such a rate can be maintained only for a short period of time. If Adam and Eve and their descendants, for example, had reproduced at such a rate, they would have filled the earth with its present population in only six to eight centuries. The relationship between doubling times and rates of growth is indicated in Table 5.1. The power of geometric multiplication is truly astounding, if it is allowed to continue for any length of time. In reality, one observes that high growth rates are maintained only for a limited number of doublings and then tend to decline. An-

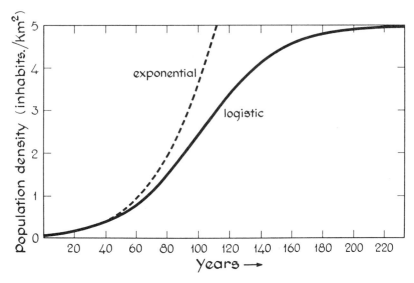

FIGURE 5.4. Two models of population growth: exponential and logistic. Under exponential growth, the population increases geometrically in size over time. Under logistic growth, the population increases in much the same way initially but slows down as a saturation level in terms of population density is approached.

other widely employed model of growth is the *logistic* one. Here the number of individuals increases according to an S-shaped curve (see figure 5.4). In the beginning, growth proceeds almost as fast as in the exponential case, but the rate of increase slows down progressively as the population size increases, and it eventually falls to zero as the population reaches a plateau.[10]

The logistic form of growth constitutes the simplest treatment in mathematical terms of the situation in which the growth process is limited. It is worth stressing that the model represents an idealized treatment of population dynamics. It does not and cannot by itself take into account phenomena that are bound to be observed in actual populations, such as changes in carrying capacity of the land due to technological innovations (e.g., the introduction of plowing in the case of agriculture), which will increase the saturation level, or events that can determine serious, often precipitous drops of population numbers, such as wars, pestilences, and famines. Thus, the number of people inhabiting a specific area may first go up, and a logistic curve may be of help in describing such a period of growth. But after it stabilizes in size, the population may undergo further increases or decreases, and more complex curves may be needed in order to characterize population trajec-

LOGISTIC GROWTH

tories over longer periods of time. The wave of advance model is intended to account for the initial spread of early farming and not long-term population trends in various regions of Europe, which are likely to require other models of a more complex character.

It will soon be clear why we are especially interested in the initial growth rate, a, in terms of a logistic treatment. In the wave of advance model, active population growth happens only at the frontier. Turning to figure 5.5, we can see that after some time there is near the origin ($x = 0$) no net increase in population size, while growth is active near the periphery. It is population growth in the frontier zone at any given time that effectively determines the expansion. Yet even within the frontier zone itself, there are variations in rates of population change. Let us examine a specific time: for example, 1,500 years after the start of the expansion (see figure 5.5). At the leading edge of the wave front, populations will experience the highest local net growth rates. As we move closer to the origin, the tempo of population change slows down and is essentially zero in those places where the population has reached saturation level. Thus, the growth rate of particular interest to us is that which obtains at the extreme frontier zone of the expansion. For estimates of growth rates appropriate for the model, we can draw on evidence from ethnography and history, but there are also ways in which archaeological evidence can be used to estimate rates more directly.

The first data deserving attention are naturally those of the earliest farmers themselves. Unfortunately, such data are extremely limited. The best data available at present come from studies of the Bandkeramik culture in central Europe. In Chapter 3 we mentioned the geographic distribution of known sites in a well-studied region near Bonn, West Germany (see figure 3.5). One hundred sixty houses were identified during excavations at eight Bandkeramik sites along a 1.3 kilometer stretch of a stream called the Merzbach.[11] Each house was probably occupied for an average of 25 years and perhaps even longer. Based upon radiocarbon determinations, the total duration of the Bandkeramik occupation in this area was about five hundred years. Ideally, we would like to know the number of people living at different points of time in order to draw a growth curve, but the data available are not sufficiently detailed. We can, however, obtain a rough estimate for the rate of growth in the following manner. If the number of houses remained the same throughout the whole period of occupation, we can calculate the total number of house-years represented by Band-

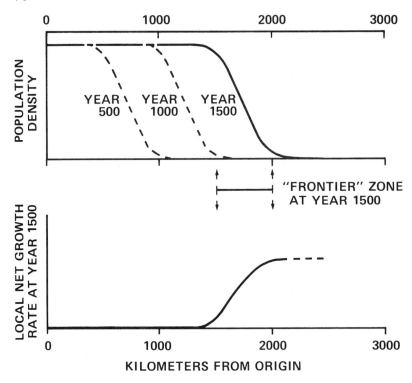

FIGURE 5.5. The relationship between local population densities and local net growth rates. At year 1,500 in the example shown here, the wave front has reached a distance of 2,000 kilometers from the origin. Changes in population density (above) are being experienced only in the "frontier" zone. Behind this zone, population densities have already reached the level of saturation, and local rates of growth have fallen to zero. Within the frontier zone itself, the net growth rate (below) is locally highest at the leading edge of the wave front and falls off as one moves toward the origin.

keramik occupation in the areas as 4,000 (160 house x 25 years = 4,000 house-years). We could also calculate the average number of houses standing on the landscape at any one time as 8.0 (i.e., 4,000 house-years/500 years of occupation = 8.0 houses). It is likely, however, that there was at the beginning only one house (representing the first household established in the area) and that others were built as the local population grew.

Let us assume that local saturation may have been reached near the middle of the period of occupation. Under these conditions, a simple calculation, presented in figure 5.6, shows that there would have been on average 11 houses standing during the second half of the occupation period. This figure assumes that one-third of all the houses were built during the first half (0-250 years) and two-

LOGISTIC GROWTH

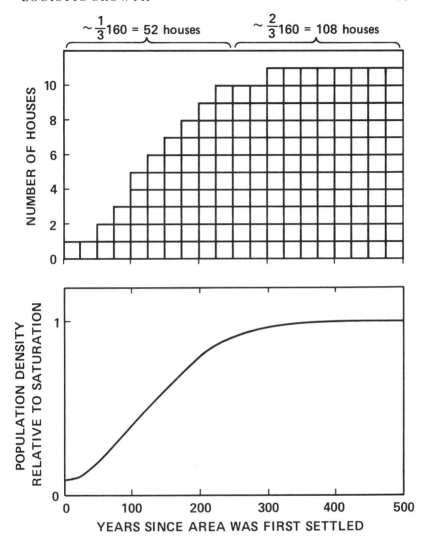

FIGURE 5.6. Heuristic analysis of population growth on the Aldenhoven Platte. Growth is measured here in terms of the number of Bandkeramik houses occupied at different times. The assumptions are made that growth occurs according to a logistic model and that each of the 160 houses identified archaeologically at Aldenhoven was occupied for 25 years.

thirds during the second half (250-500 years). According to a logistic form of growth, we can estimate the growth rate a as 1.9%.[12] The value of 25 years taken here for the length of time that a house was inhabited is probably on the conservative side. The estimates should obviously be regarded as rough ones, but the analysis is

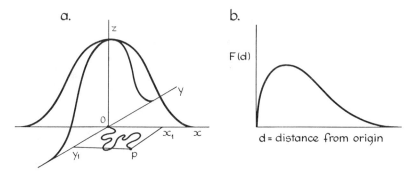

FIGURE 5.7. The treatment of migratory activity in Fisher's model. Movement is considered to be random in direction, and the frequency distribution of distances moved is considered to correspond to a bi-dimensional Gaussian surface. On the left, the value of z gives the frequency of individuals expected to reach a point P (x_1, y_1) at a given time. On the right, this distribution surface has been transformed into a distribution curve where the distances moved are given by $d = \sqrt{(x^2 + y^2)}$.

useful as a heuristic exercise in showing how rates of growth can be estimated from archaeological evidence. It should be possible in the near future to obtain estimates of this kind in other areas of Europe and to refine the actual approach to estimation.

5.4 MIGRATORY ACTIVITY

The term "migratory activity," rather than "migration," is intentionally used here to indicate processes involving the movement or relocation of individuals within a population where the distances moved are usually quite short. One form that such activity often takes is change of residence of either husband or wife at the time of marriage. In contrast, our concept of the term migration usually carries with it the notion of long-distance movements involving larger groups of people. In Fisher's model, the treatment of the migratory component follows the classical random one for diffusionary processes. This treatment, which is comparable to Brownian movement for molecules, is mathematically the most tractable formulation to consider. Migratory activity is regarded as taking place continuously in time and space, with the frequency distribution of migration distances conforming to a bi-dimensional "normal," or Gaussian, surface (see figure 5.7). The values of the z axis of the figure give the probability that an individual located at point 0 at the beginning will reach a point having the coordinates (x_1, y_1) after a given time. It is usually more convenient for this distribution surface to be transformed into a distribution curve, with the distance from the origin being measured in terms of the standard Pythagorean formula: $d = \sqrt{x_1^2 + y_1^2}$. The shapes of the Gaussian surface and the transformed distribution curve are both described by m, the mean square distance of movement per unit time.

It is evident in the case of early neolithic populations in Europe that "movement" did not take place continuously in space and time and that a discrete treatment would be more realistic. Individuals move from one settlement, which has a fixed location in space, to another settlement. With respect to time, movement occurs at intervals or periodically rather than continuously. Simulation studies provide one way of checking the possibility of a major divergence between what is observed under a discrete treatment and what is expected to occur under the model and its continuous formulation. Studies in which the same parametric values for m and a are used in the two different treatments indicate that a reasonable approximation is maintained. In simulation work undertaken by Skellam, it was found that a discrete treatment tends to "blunt" the leading edge of the wave front, thus slowing down somewhat the velocity of the expansion.[13]

From a conceptual point of view, it is important to think of migratory activity in terms of settlement patterns. The actual pattern of expansion is generated to a large extent by the relocation of farming households and settlements to adjacent areas where the land has not been cultivated previously. How sites are arrayed over the landscape conditions in a general way such things as the relative frequency of movements and the size of migration distances. We can obtain an intuitive sense of this if we contrast two kinds of settlement systems: (a) a dispersed one with many small sites located at relatively short distances from one another, and (b) a more aggregated system where many families live at any one site and the average distance between nearby settlements would be longer. In the first case, population growth will be manifested primarily by an increase in the number of new settlements. A person living at such a site is unlikely to find a marriage partner at the same site (i.e., a person of an appropriate age, belonging to the opposite sex, not having a consanguineous relationship, and not already married). Marriage will usually require that one or both partners change their residence. Although the level of migratory activity will be high in such a case, most of the movement will probably involve quite short distances. In the second settlement system, population growth can take the form of both an increase in the number of households at a settlement and an increase in the number of settlements on the landscape. At a large settlement, it may well be easier to find a marriage partner within the same settlement, and the level of migratory activity would consequently be lower. But when movement does occur, it is likely to be on average over a longer distance. There is a good chance that the shapes of the frequency distributions of migration distances will differ between the two systems. In the context of the spread of early farming in Europe, the settlement patterns of the Bandkeramik culture of central Europe would seem to reflect more the former system, whereas the mound sites of early neolithic cultures in Greece and the Balkans would come closer to the latter. Of course, it is clearly possible to imagine settlement systems other than the two put forward here as a means of illustrating some of the relationships between settlement patterns and migratory activity.

Direct estimates of m, the mean square distance of movement per unit time, are difficult to obtain from the archaeological record. We can gain some idea of the size of m from ethnographic studies. An upper bound for this distance is provided by the "exploration range" of an individual, or the average distance from a person's birthplace to the other places known to that person.[14] In the case of African pygmies, a hunting and gathering people, this distance is 88 kilometers for adult males and 57 kilometers for adult females. The exploration range for most agricultural groups is likely to be substantially shorter. Moreover, this measure may well overestimate the distance to actual places of residence during an individual's lifetime. A measure that may give a better picture of migratory activity is the distance between the birthplaces of husbands and wives. In figure 5.8 the distributions of such distances are shown for three modern populations. Note here that the distributions tend to have a somewhat different shape than the one expected under the formal model (see figure 5.7). The population of hunters and gatherers, again African pygmies, exhibits much larger mating distances than the two populations of farmers. The average mating distance for African farmers (11.9 kilometers) is similar to that of European farmers (13.5 kilometers). But when the mean square distance (m) is considered, the European farmers (1,049 square kilometers per generation) appear to be more mobile than their African counterparts (335 square kilometers per generation). The low value of the latter may be owing to characteristics of the tribe involved; the Issongos of the Central African Republic occupy a very limited area, and few marriages occur with nearby tribes.

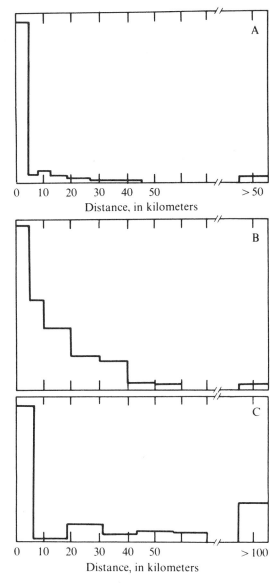

FIGURE 5.8. Migratory activity as reflected by distances between the birthplaces of husbands and wives. The three frequency distributions are for the following modern populations (Cavalli-Sforza and Bodmer 1971: 414): (A) Upper Parma Valley (Italy), (B) Issongos (the Central African Republic), and (C) Babinga pygmies (the Central African Republic).

Estimates of migratory activity can also be calculated for the Majangir, a population of shifting agriculturists in Ethiopia, on the basis of data on settlement relocation distances.[15] The values of m, which range between 1,100 and 2,100 square kilometers per generation, are somewhat larger than those obtained from mating distances. In the next section, values that vary between 300 and 2,000 square kilometers per generation will be used for the migratory component in developing a test of the wave of advance model.

5.5 An Initial Test

Each of the two main components of the wave of advance model has been examined in a preceding section of this chapter. As an initial test of the model, we can ask the following basic question: Does the observed rate of advance, estimated to be one kilometer per year, agree with the rate expected on the basis of the model, and, specifically, does it agree with what we know about rates of growth and migratory activity? If there is no agreement between the two, then we can reject the model as a candidate for explaining the spread. The strategy adopted for evaluating a hypothesis or model is a refutationist one. According to the model, the expected rate of advance r is twice the square root of the product of the initial growth rate, a, and the measure of migratory activity, m (the mean square distance of movement per unit time). From the discussion of the two components, we have some idea of the range of acceptable values for the two parameters.

- a (growth rate): The values here are in the range from 0.6% to 3% per year, with an estimate around 1% per year perhaps the most likely. Values above 3% are practically impossible, while those below 0.6% are not completely excluded.

- m (migration rate): The values most likely fall in the range between 300 and 2,000 square kilometers per generation.[16]

Each curve in figure 5.9 shows all possible pairs of values for m and a that are associated with a given rate of advance expected under the model.[17] The observed rate of advance is shown by a thicker curve, on which we can read that the lowest value mentioned above for the growth rate, 0.6% per year, would be associated with a value of 1,000 square kilometers per generation for migratory activity, which falls in the middle of the range of likely values. The

AN INITIAL TEST

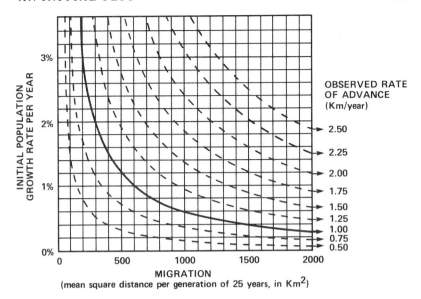

FIGURE 5.9. An evaluation of the wave of advance model. The curves indicate the rate of advance that would obtain under various combinations of values for M (rates of migratory activity) and a (rates of population growth). The heavier curve represents the values that in combination would produce a rate of advance of 1 kilometer per year.

maximum growth rate, 3% per year, has a corresponding value of 300 square kilometers per generation for migratory activity. Associated with the maximum value for migratory activity—2,000 square kilometers per generation—would be a growth rate of 0.3% per year, which falls below the range of values considered to be likely for the initial rate in a logistic form of growth. The conclusion to be drawn is that the rates of advance predicted by the model, when reasonable values for the rates of growth and migratory activity are employed, are essentially compatible with the observed rate of advance. This conclusion tells us that the postulated mechanism for the spread of early farming in Europe is in agreement with lines of evidence that are available. It does not tell us that the wave of advance model is necessarily the correct or the only explanation. The model has simply managed to survive an initial test. It is important at this stage to consider other kinds of tests or predictions that can be employed to evaluate the demic hypothesis. We shall see that the wave of advance model has major implications for the genetic composition of populations in Europe and that the investigation of genetic patterns may represent one of the more promising ways to test the demic hypothesis.

5.6 Genetic Implications of the Model

At the end of Chapter 4 we discussed two contrasting ways of explaining the spread of early farming in Europe: the cultural mode of diffusion of the idea of farming and the demic mode. As noted, the two modes of explanation need not be regarded as mutually exclusive but can operate alongside of each other. However, the two modes of explanation do give rise to totally different expectations with regard to the geographic patterns of gene frequencies in human populations. It is useful to consider the consequences of three possible models: a purely demic model, a purely cultural model, and a mixed model.

In order to illustrate the different patterns that arise in terms of human population genetics, a highly simplified treatment of the problem will be introduced here. Let us assume that prior to the origins of agriculture, both Europe and southwestern Asia were inhabited by populations of hunters and gatherers that exchanged migrants only in a very limited way. Genetic differentiation, as we shall see in the next chapter, can develop over time between populations if they live essentially in isolation from one another. Let us further assume that differentiation is complete with respect to a hypothetical gene, so that all of the people living in Europe have a form of this gene different from that found in southwest Asia. If the gene exists only in form A in Europe and only in form A' in Asia, we would have the following situation:

	W. Europe	E. Europe	S. W. Asia
% of A genes	100%	100%	0%
% of A' genes	0%	0%	100%

Here Europe is divided into western and eastern halves for reasons that will become clear below.

Suppose that the diffusion of agriculture is entirely cultural. Then no changes will occur in the geographic pattern of the hypothetical gene: after the introduction of farming in Europe, the gene frequencies in the respective regions will remain the same as above. On the other hand, if we suppose that the diffusion is entirely demic (e.g., farmers from southwestern Asia fully replace the hunter-gatherers living in Europe), then after the spread of agri-

GENETIC IMPLICATIONS OF THE MODEL

culture has run its course, the geographic pattern of gene frequencies will change to the following:

	W. Europe	E. Europe	S.W. Asia
% of A genes	0%	0%	0%
% of A' genes	100%	100%	100%

In other words, the gene frequencies now seen in Europe correspond to those initially present in southwest Asia. In terms of genes, it is clear that the expectations under the cultural and demic hypotheses are sharply different.

Under the mixed model—a combination of demic and cultural diffusion—intermarriage is considered to occur between spreading agriculturalists and local populations of hunters and gatherers in different parts of Europe. In Chapter 7 we shall give a more precise treatment of the factors and assumptions that are involved under this mixed hypothesis. Again for heuristic purposes, it is useful to indicate the kind of pattern that would result from such an interaction. Let us assume for the sake of illustration that, starting from the initial situation described above, the process takes place in two steps only: first, migration from southwest Asia to eastern Europe and, second, from eastern Europe to western Europe. After the first step, the gene pool in eastern Europe will comprise a 50:50 mixture of people from eastern Europe and those from southwest Asia. The gene frequencies of populations in southwest Asia itself and western Europe will remain unchanged.

	W. Europe	E. Europe	S.W. Asia
Gene A	100%	50%	0%
Gene A'	0%	50%	100%

After the second step, populations in western Europe will consist of a 50:50 mixture of the gene pool of western Europe and that of eastern Europe (i.e., the gene pool established in eastern Europe during the first step):

	W. Europe	E. Europe	S.W. Asia
Gene A	75%	50%	0%
Gene A'	25%	50%	100%

During the second step there is considered to be no further migration from southwest Asia to eastern Europe. We can see that the mixed model has established a cline—a gradient of gene frequencies running from east to west—along the main axis of the expansion. A fuller treatment of the mixed model in terms of population interactions and expected genetic patterns will be developed in Chapter 7.

One prerequisite for the emergence of such patterns is that there was, prior to the origins of agriculture, sufficient differentiation between populations in southwest Asia and Europe in terms of their gene frequencies. Ideally, one would like to use genetic information that derives directly from the time period of the spread of agriculture. Genetic markers, unfortunately, cannot be obtained at the present time from skeletal material. They can be investigated only in living populations, which introduces further possible uncertainties. One obvious source of complication is that considerable time has elapsed since the spread of agriculture; various events, including subsequent migrations, may have occurred in the meantime, generating complex patterns. In addition, processes such as natural selection may have transformed or obliterated the patterns expected under the model of demic diffusion. It was, therefore, with some trepidation that we started an analysis of the vast body of genetic data on living populations in Europe.

CHAPTER 6

THE ANALYSIS OF GENES

6.1 EXPECTATIONS FOR THE GEOGRAPHIC DISTRIBUTION OF GENES

For the geneticist, it is not unreasonable to speculate that the origins of agriculture, perhaps more than any other cultural change, may have had major consequences for the evolution of genes and their patterns of geographic distribution. In the last chapter we saw that if agriculture spread by means of cultural diffusion, it should leave the previous gene distribution intact, whereas a completely demic diffusion would lead to the replacement of populations in Europe—and therefore their genes—with those from southwestern Asia. A mixed cultural-demic diffusion should generate a gradient (a *cline*, using a biological term) in the direction of the migration: that is, the genes of the original farmers would decrease proportionally as one proceeds from southwestern Asia toward Europe. In the last section of the preceding chapter, Europe was, for simplicity, divided into two major subregions. This treatment offers only the roughest opportunity for describing a gradient. More refined models will be presented later in this chapter.

The ideal situation for distinguishing among different expectations is obviously to find a given gene at 100% among all populations in Europe and at 0% among those in southwestern Asia. Such a clear difference is unlikely to be observed, since the genetic differentiation between Europe and southwestern Asia prior to the spread of agriculture was unlikely to be so great for any gene. When differentiation is incomplete, it will be useful to sum the evidence from a number of genes. This approach will be an essential part of our procedure, since we want to use as much of the available information as possible in the evaluation of our models. Patterns of the geographic distribution of genes, which are hard to understand if taken one by one, can by means of synthetic treatment help in producing maps that permit us to infer the histories of past population movements, thus making sense out of an accumulation of apparently unordered facts. Before we describe the actual methods employed in the synthetic treatment, we want to illustrate the geographic distribution of several genes and give a short summary of the evolutionary forces at work. It will become clear that the geographic distribution of a gene is influenced not only by the process of natural selection but also by population movements and histories.

6.2 The Rh Gene

In his study of the geographic distribution of the Rh blood groups, A. E. Mourant noted that the Rh negative gene (Rh−), which is found almost exclusively in Caucasians, has its highest frequency among the Basques, who live in or near the valleys of the Pyrenees at the boundary between France and Spain.[1] Basques speak a language that is quite different from the Indo-European languages spoken in the rest of Europe. Mourant hypothesized that the Basques represent the descendants of the oldest inhabitants of Europe, who mixed at a later date with immigrants from outside the region. Thus, Mourant's hypothesis requires that ancient Europeans were predominantly, if not exclusively, Rh− and that later immigrants were predominantly, if not exclusively, Rh positive (Rh+). It does not place the later immigrants in any particular place, since all non-Europeans are mostly, if not exclusively, Rh+. Nor does it indicate whether or not the Basques at the time of the admixture were hunter-gatherers or farmers. But it is reasonable to ask if the immigrants could have been early farmers reaching the areas as part of the spread of agriculture from the eastern Mediterranean.

The discovery of the Rh gene takes us back to 1939, when P. Levine and R. Stetson described a case of fetal death accompanied by signs of profound anemia of the fetus.[2] Other clinical cases were subsequently found, and about one in 200 births among Caucasians turned out to be so affected. The mother's blood was shown to contain substances or antibodies that reacted in a characteristic way (that is, by clumping) with the red blood cells of the fetus. These antibodies were similarly active against the red blood cells of her husband but not against her own red cells. It was found that such antibodies reacted with the red blood cells of some 84% of all white American people tested (called Rh positive) but not with the other 16% (called Rh negative). Among non-European populations, the Rh+ compounds were found to be present in the vast majority of people (almost 100%). It was independently discovered that the same substances are in the blood of many monkeys and apes: in fact, the name Rh comes from the initials of the rhesus monkey. Some compound was present on the surface of the red blood cells of the fetus, the father, and the majority (84%) of white people, but absent in the mother of the fetus and in 16% of all Caucasians capable of reacting with the antibody. The medical problems raised by Rh, once serious, have been almost entirely solved in the last few years by ingenious methods of monitoring and prevention.

As in the case of almost all other genes, we receive two copies of the Rh gene, one from each parent. A person is Rh negative if, and only if, he or she receives from *both* parents a slightly altered "gene" that, for our present purposes, can be called r or the Rh negative "gene."[3] If we look at a map showing the frequency of the Rh− gene (figure 6.1), we find that the highest percentages of the gene of Rh negatives occur not only near the Pyrenees but also in northern Europe—approximately the periphery of the expansion of early farming. In central Europe the frequencies range between 12% and 15%. In southern Italy, Greece, Turkey, and the Near East, they are between 9% and 12%. Toward the south (Arabia and Africa) and the east (India), the percentage of Rh negatives is still lower. For clarity, we should add that the frequency of the Rh− *genes* and that of Rh− *individuals* are not the same thing. The latter frequency is always lower.[4] Thus, the overall picture is not in disagreement with the idea that farmers spread from southwestern Asia. Prior to the spread of farming, populations in southwestern Asia had a low proportion of Rh negatives, while populations in Europe presumably had high Rh− values. This statement may be an oversimplification, but it is not impossible that the Rh gene may represent the ideal case for studying the admixture between populations that we hypothesized at the end of the last chapter. However, some complications relating to the natural selection to which Rh genes are probably exposed make this simple assumption less than acceptable.

6.3 Mechanisms of Evolutionary Change

Attempts to interpret the geographic distribution of the Rh− gene raise a number of basic questions. Why are there many forms of a gene such as the Rh? Why do the frequencies differ from one place to the next? Do they change over time? With regard to the first question, all of the different forms of a gene arise by means of a process called *mutation*. As is well known, reproductive cells contain in their DNA a complete blueprint—that is, a complete set of genes— for reproducing a full organism. Thus, each sperm or egg contains

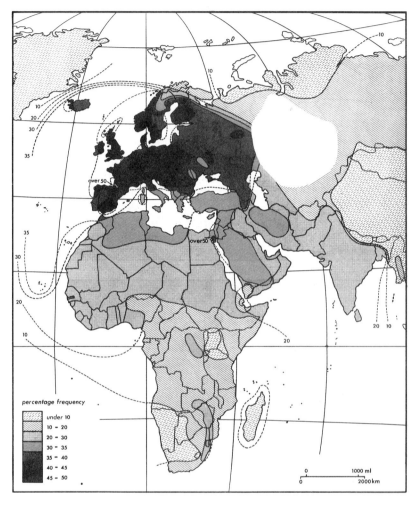

FIGURE 6.1. Map of the Rh− gene in the populations of Europe, Africa, and western Asia (Mourant et al. 1976: fig. 22).

one copy of the Rh "gene." The many billions of sperm cells produced in the life of a male ordinarily represent perfect copies of the genes of the individual, but very occasionally a small reproduction error produces a gene slightly different from the one being copied in the parent. This error is known as a mutation, and the "new" gene thus formed is called a mutant. Similarly, mutations can happen in eggs. If a mutant gene is passed to an offspring, that individual may show a very small, but significant, difference from the parental type.

At the DNA level, the difference between the mutant gene and the ancestral one is usually very small. The mutant gene will ordinarily keep making the same substance, perhaps with only very slight modification. That modification, however, may be detected by sensitive biochemical reagents. This is the case for the various forms of the Rh positive gene: R_1, R_2, R_o. It is also possible that the mutation may be more important and destroy entirely the activity of the Rh gene. This is what may have happened in the case of the Rh− gene. There may be no true disadvantage associated directly with this loss. It is only when an Rh− mother has an Rh+ child that trouble begins to arise.

Mutations are rare: a gene happens to be copied incorrectly perhaps once in a million times. But so many new individuals are formed every generation and so many generations have existed in the evolution of living beings that many mutations have the chance to occur. For any one human gene, there is formed each generation (approximately every thirty years in modern populations) some ten billion copies destined to become new adults—two copies of a given gene for each of the five billion individuals living on the planet today. Of the ten billion new Rh genes produced during a generation—each potentially contributing to the makeup of a new individual in the next generation—one in a million may experience mutation. Thus, several thousand mutations of the Rh gene are produced every thirty years. Each may be slightly different, and perhaps some of them entirely new.

One may wonder, then, if any two genes can ever be the same, given this variation occurring every generation. In fact, most new mutations are lost, many of them because of sheer chance. Suppose that a new mutant gene is passed to a child by one parent, say the father; the child will also receive another Rh gene from the mother. When the offspring carrying such a mutant gene of paternal origin has children, he or she will pass on the mutant gene with a probability of only 0.5. There is an equal chance that the gene passed on will be the one derived from the other parent. If this individual has only one child, the chance that the mutant gene will be lost is 50%. Even if this individual has more than one child, the chance that none of them will carry the mutant gene is not negligible. This chance process, it turns out, causes almost all new mutants to be lost eventually. Occasionally, a mutant may, after a very long number of generations, reach complete success and become the only form present in a population. Chance or genetic drift is the only process of evolutionary change that is considered to be operating so far in this discussion. Under chance alone, a single mutant may come to dominate the whole population, but it will do so only rarely and, we may add, very slowly.

The odds that a new mutant either will be lost or will sweep through a population are strongly influenced by any effect that the mutant gene may have on an individual's ability to survive and reproduce. Suppose that the new mutant gene confers an important advantage to its carrier, who will thus have a better opportunity than otherwise of contributing offspring to the next generation. The new gene will then not only have a greater chance of spreading through the population but may do so in a relatively short time, as indicated by the curve of figure 6.2. This is *natural selection* in favor of an advantageous mutant. The rate at which the mutant will increase in comparison with other forms of a given gene in a population will depend on how advantageous it is and also on how large the population is. In a small population of 1,000 individuals, for example, a new gene with an advantage of 10% over the old form may come to permeate the population within as few as fifty generations. A *selective advantage* of 10% means that an individual carrying the new gene will have, on average, 10% more children that survive to adulthood than other gene forms. Such a value would represent an unusually high selective advantage. Our knowledge of the selective advantage of genes is still very limited, and most selection processes are believed to be actually much slower.

Naturally, the reverse can also happen: the new gene may be less advantageous to its carrier than the old gene. In this case the new gene has almost no chance of ever being represented by a substantial number of individuals in a population. The majority of new mutants happen to be of this deleterious type. The more disadvantageous a new mutant, the more likely its rapid elimination. Thus, natural selection assures that individuals of a population are as "fit" as possible. "Fitness" in population genetics refers not to the strength of an individual but specifically to the relative number of descendants that a given genetic type contributes to the next generation. The process of selection is said to be "natural" in the sense that the environment in which a population lives effectively determines which genes increase and which decrease the capacity of an individual to survive and reproduce. Thus, animals that live in caves, screened from all sources of light, can become blind because there is little or no disadvantage attached to being blind. But,

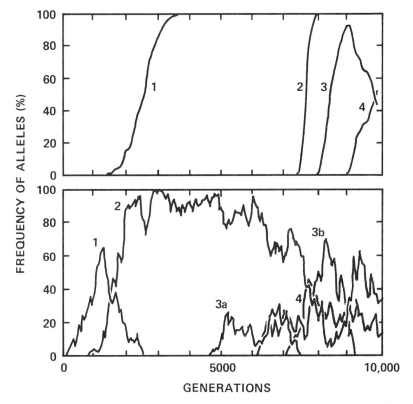

FIGURE 6.2. Comparative simulations of two populations. In the top figure, four mutations have arisen that have a selective advantage: the first 0.36% higher than normal, the second 0.50% higher, the third 2.20% higher, and the fourth 2.40% higher. The population was made up of 2,000 individuals, the mutation rate was 1 in 10,000, and each mutant had a different selective advantage. Only mutants with sufficiently high advantage have reached fixation. In the lower figure, all mutations that arise are selectively neutral. Only one of them ever went close to fixation, and the frequency of each mutation shows considerable oscillation in time (from Bodmer and Cavalli-Sforza 1971).

of course, a mutant producing blindness would hardly ever survive in an environment in which light is present.

Since 1968 researchers have been debating the proportion of mutations that are or are not selectively *neutral*—that is, neither advantageous nor disadvantageous. It seems quite clear that at least *some* are, if not neutral, at least so close to neutral that their advantage or disadvantage, if any, is undetectable. The theory that *many* of the mutations commonly studied in the laboratory are selectively neutral was first advanced about ten years ago by Motoo

Kimura, a Japanese geneticist, and almost simultaneously by J. L. King and T. A. Jukes in California.[5] Considerable controversy has resulted; biologists strongly adhere to the idea that natural selection is the only factor permitting adaptation. Although correct, this does not necessarily mean that all evolution is directed by natural selection and, hence, adaptive. Some evolution may be of random nature. Kimura's hypothesis has recently gained favor. However, the full answer to this question will come only when there is enough evidence to state: (1) what proportion of mutants behaves as truly adaptive (i.e., favored by natural selection), (2) what proportion is practically uninfluenced by it (i.e., truly neutral), and (3) what proportion is disadvantageous. Scientists generally agree that most mutants, perhaps 90% or more, are neither adaptive nor neutral but truly deleterious. These are rapidly eliminated by natural selection and do not call for further consideration here since they do not contribute in an important way to the variation that we observe in human populations.

This general problem of neutrality versus selection is important for our purposes. Let us imagine, again probably oversimplifying, that prior to agriculture there are people in Europe who are 100% Rh−, while those living in southwestern Asia are 0% Rh−. Then the proportion of Rh− genes among modern Basques (that is, 55%; see note 4) would tell us that they originated from a mixture of 55 European hunter-gatherers and 45 early farmers from the Near East. To be valid, this reasoning would require not only that the original frequencies of 100% and 0%, respectively, are correct, but also that no other circumstances caused change in the gene frequencies after the admixture: in other words, that the Rh positive and negative types are selectively neutral. But the Rh gene would seem to be far from ideal for such a study, since we know that Rh+ progeny of Rh− mothers have a high chance of dying. As it happens, however, the particular mechanism of natural selection in the Rh case is perhaps not at odds with our aims. So far as we know, neither the Rh+ gene nor the Rh− gene is necessarily "good" or "bad" from the point of view of natural selection. The Rh+ fetus that dies because of Rh disease contains one Rh+ gene from the father and one Rh− gene from the mother, and thus, in the present case, natural selection is relatively impartial. A theoretical analysis shows that the Rh+ gene eventually will be eliminated in the presence of a majority (more than 50%) of Rh− genes. On the other hand, the Rh− gene will be eliminated in the presence of more than 50% of Rh+ genes. In practice, the whole process is

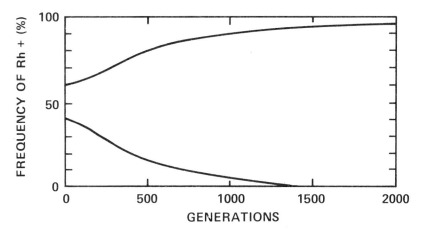

FIGURE 6.3. Theoretical selection curves for the Rh gene under effect of selection due to Rh hemolytic disease. Rh+ gene frequencies greater than 50% will tend to eliminate the Rh− gene, and frequencies of the Rh+ gene below 50% will tend to eliminate the Rh+ gene. Intensity of selection is set at the level currently observed. The evolutionary time is given on the abscissa as number of generations (after Cavalli-Sforza and Bodmer 1971: 201).

relatively slow, as figure 6.3 shows. Nevertheless, the rate must be taken into account if we want to estimate the proportions of admixture correctly. At the same time, the Rh marker may be especially revealing, since under certain conditions it reacts to the introduction of foreign genes by tending to remain in status quo longer than other genes. If a population that is predominantly Rh− in character receives Rh+ immigrants, it will slowly eliminate them; hence the population's original composition may change more slowly than in the case of a gene that does not have this specific property. This may in fact be the reason why the Rh gene stands out among many others in pointing to an admixture between early European and immigrant populations. It is clearly of interest to see whether or not similar patterns are revealed by other genes.

6.4 Evidence from Other Genes

Ideally, we would like to be able to base our analysis on genes that are not subject to natural selection. If this is not possible, it is preferable to use genes for which there is no evidence for strong selection; otherwise, we can expect to learn more about selective processes than about population histories. It is also easy to understand why we would prefer genes that exhibited considerable differences in frequency between Europe and the Near East prior to the spread of early farming. However, as we have already mentioned, we do not possess information on gene frequencies in the remote past. We can only select genes that *today* show variation in the part of the world we are studying and hope that these differences are not due to subsequent events.

It may seem that we are asking for too much. But we have the advantage of currently possessing in Europe a substantial amount of information on a large number of genes, some of which are represented by many different forms. The more numerous the different forms of a gene (as in the case of Rh and HLA), the more informative that gene. For the Rh system, we have so far considered only the negative form; there are several others. We should emphasize that we are basing our conclusions on the results of an enormous amount of work done by many laboratories. The blood of several hundred individuals must be sampled in order to describe just one population. For a good description of a region such as Europe, one needs data of this kind for at least fifty populations, preferably well spaced over the area to be studied. The tests to be carried out on the blood samples are demanding, numerous, and sometimes expensive. Despite the difficulties, there are today more than four thousand publications giving data on gene frequencies of human populations. But relatively abundant data are available for only those few genetic markers that have been known for some time and that have recognized biomedical importance. Observations are almost without exception more numerous for Europe. They were generally scarce in Africa, southwestern Asia, and parts of the Soviet Union.

The gene that has been studied most widely is the ABO blood group system. Knowledge of ABO is necessary for blood transfusions, and ABO was the first gene to be studied in human populations. As early as 1917, geographic variations in gene frequencies were observed, and it was suggested that this system could be used for tracing ethnic origins. Of the three major forms of the gene (i.e., A, B, and O), B in particular shows considerable differences among populations. The B gene has lower frequencies in Europe than in Asia or Africa and happens to be almost entirely absent in American natives. The patterns of variation in Europe for the three types are shown in figures 6.4–6.6. Each of the three maps presents a different pattern: A exhibits a gradient of declining frequencies running from the northeast toward the southwest, while B shows

EVIDENCE FROM OTHER GENES 95

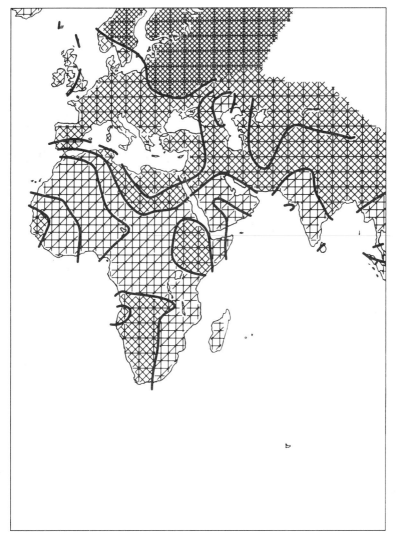

FIGURE 6.4. Map of the gene frequency of blood group A in European populations (from Bodmer and Cavalli-Sforza 1976).

a gradient running from east to west across Europe. The pattern for O is roughly similar to that of A, but in the opposite direction. Naturally, the three frequencies are correlated: if one increases sharply, the others—at least one of the two—must decrease, since their percentage values have to sum to 100% in any location. Yet

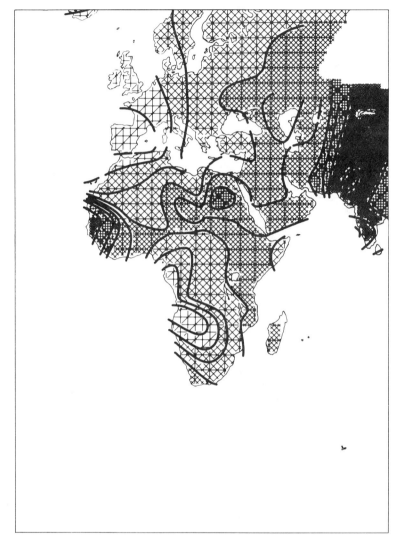

FIGURE 6.5. Map of the gene frequency of blood group B in European populations (from Bodmer and Cavalli-Sforza 1976).

none of the three types (A,B,O) gives results in agreement with our expectations. If the gradients or clines are the consequence of migrations, the distribution of type B would indicate that the migrants came from central Asia and not southwestern Asia. It has in fact been suggested that blood group B was common among

EVIDENCE FROM OTHER GENES

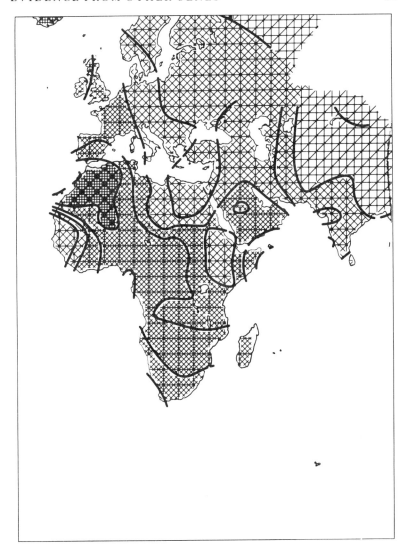

FIGURE 6.6. Map of the gene frequency of blood group O in European populations (from Bodmer and Cavalli-Sforza 1976).

central Asian people, and there may be some evidence for population movements of prehistoric and historic date from this part of the world.

There might be reasons other than migrations for some of the variation of ABO blood groups. For example, the north-south gradient of type A reflects variations connected with climate. Claims have also been made that individuals belonging to various groups—in some cases A individuals, in others B or O—are slightly more resistant to certain infectious diseases. Although most of these claims are ill-founded, there is certainly evidence to support the idea that certain blood groups are influenced by natural selection. With regard to many of their genes, Ashkenazi Jews, who have lived in central and eastern Europe for more than a thousand years, are still similar to populations from the Near East, the area from which they came some two thousand years ago. On the other hand, with respect to some genes, in particular ABO, they are more similar to European populations. The simplest interpretation is that they have had only very minor exchange (i.e., cross-marriage) with their European neighbors and that the ABO frequencies vary, however slightly, in the same direction as those of neighboring populations probably as a response to local selection. If this is true, ABO blood groups may be less good indicators of ancestral origin than was once thought.

Although many different gene systems have been identified, not all have been studied extensively enough so that good maps can be drawn. Only ten systems, each of which has several different gene forms, have been investigated in sufficient detail for our purposes. Of special interest are the HLA genes, which have attracted much attention in the last ten years for purposes of organ transplants and also as tracers of certain rare genetic diseases. Although they also must be under selection, the HLA genes seem to be relatively stable from an evolutionary point of view, as can be inferred from the fact that they vary from place to place less conspicuously than almost all other genes. Taking the system as a whole, some 60 detectably different forms exist. Because of the large number of different forms known, the HLA system offers, potentially at least, some of the best information for studying geographic patterns of genetic variation in European populations. It is almost as informative as all the other available gene systems taken together. Rather than showing geographic maps of the 30 or so different HLA forms that have been studied on a sufficiently large scale so that gene frequency maps can be constructed, we have selected one of them (see figure 6.7) because it shows agreement with the Rh distribution and, therefore, with the idea of a spread of early farmers from southwestern Asia. Another map (figure 6.8) shows more agreement instead with the geographic pattern of type A of the ABO system. From looking at individual gene maps, which show considerable variation from one to another, it is difficult to obtain

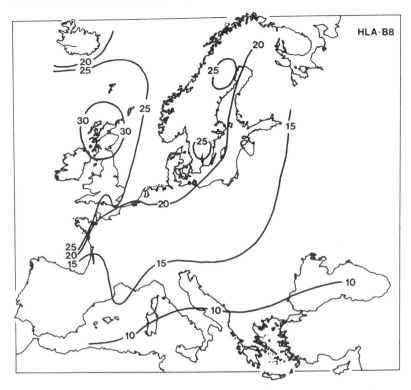

FIGURE 6.7. Map of the frequency of the HLA-B8 allele in European populations (Ryder et al. 1978: fig. 1).

a clear sense of overall genetic trends among populations in Europe. We need a more synthetic approach to the spatial analysis of genetic information.

6.5 TOWARD A SYNTHETIC VIEW

Ten gene systems (ABO, Rh, MNS, Le, Fy, Hp, AP, PGM, HLA-A, HLA-B) have been studied for a large enough number of populations in Europe to provide data of the kind needed for making gene maps. We decided to exclude genes that are known to be indicative of local environments and in particular those that determine diseases such as thalassemia and glucose 6 phosphodehydrogenase deficiency, which are associated with areas of the Mediterranean traditionally (i.e., through the Second World War) having

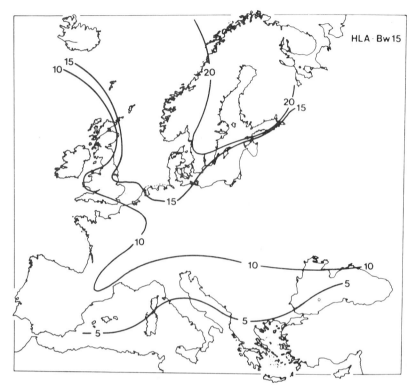

FIGURE 6.8. Map of the frequency of the HLA-Bw15 allele in European populations (Ryder et al. 1978: fig. 2).

a high incidence of malaria. The ten gene systems include a total of 39 different gene forms that can be mapped individually. The need to synthesize somehow the information from 39 genes represents a considerable challenge. Since the problem had not previously been explored in any depth, there was no ready-made method of analysis to which we could turn.

A first attempt at developing a method of analysis, undertaken in collaboration with L. Sgaramella-Zonta, involved making comparisons between one population and another in terms of geographic distance and genetic "distance" (that is, an index summarizing the differences in genetic composition of the two populations for all known genes).[6] The further away that a population is from the center of dispersal geographically, we thought, the greater should be its genetic distance. If one computed (for all populations on a map) these two distances in reference to a population in southwestern Asia representing the center of dispersal, there should be a high correlation between the two measures. The values of the correlation should decrease as the location of the reference population is moved away from the center of the spread. In simplified test situations the method was found to give the right answer.

But in the real world, many complications arise. The forces that generate patterns of variation in gene frequencies are highly unpredictable. The action of genetic drift is typically random in direction, and magnitude is predictable only in terms of probability. Natural selection causes variation in specific directions, but these directions cannot easily be predicted on the basis of our present knowledge. Thus, just prior to the spread of early farming, each gene might have presented a unique and complex geographic pattern. Gene 1, for example, might have been common in Lapland, rare in the Ukraine, intermediate in western Europe, and high again in Turkey. Gene 2 might have presented an entirely different pattern, and so on for other genes.

In order to test the method, a simulation study was undertaken based on what is known of the process of evolution. For purposes of this study we assumed that only the least predictable force of evolution—namely, genetic drift—was responsible for genetic differentiation between populations. A more detailed account of this study and of computer simulations in general will be given in Chapter 7. When the method was applied to the simulated spread of early farming, it seemed to work reasonably well, provided that enough genes were included. Apparently, the law of large numbers has to be allowed to operate if we are to see through the stochastic fluctuations of individual genes.

However, when the method was applied to the real data from Europe, the results were ambiguous: the Near East *could* have been the center of origin, but other areas could not be excluded. In particular, Lapland and England could have been the centers with almost equal likelihood according to the method. These two latter results are unlikely and seem to point to a weakness in the method. It soon became clear that the method would easily give ambiguous results in those cases where the source of the spread is located at the periphery of the area being studied. Unfortunately, not much data is available for regions south and east of the areas in southwestern Asia where food production began. It was necessary to find a method that would retain its analytical power when the center of dispersal was situated near the periphery of the area under study.

After several attempts, we selected a method that could overcome this difficulty.[7]

6.6 Principal Components Analysis and Synthetic Maps

A solution to the problem of the source of spread was finally achieved by turning to an existing statistical method—principal components analysis—which we applied in a somewhat novel way. In order to consider genetic patterns in a synthetic manner, the original data on 39 genes for each population were replaced by a few summary index numbers. In turn, geographic maps based on individual index numbers were constructed. The indices used are the "principal components," which represent essentially "weighted" means of all of the gene frequencies. In order to gain some idea of how the weights are obtained, suppose that x_A is the frequency for blood group gene A at a given place; x_B is the frequency of gene B at the same place; and x_{Rh-} that of gene $Rh-$ at the same place; and so on. The index to be constructed takes the form:

$$X = ax_A + bx_B + \ldots + px_{Rh-}$$

The quantity X is easily calculated if we can determine in an appropriate way the "weights" $a, b, \ldots$, respectively associated with each gene type (i.e., x_A, x_B, and so forth). Thus, X is a new quantity that somehow summarizes all of the gene frequencies at a given place and takes a specific value for each location on the map.

An example of the way in which such quantities synthesize information may be useful. Suppose that we observe the following gene frequencies for four populations:

Location	Gene A (%)	Gene B (%)	Gene C (%)
Rome	29	8	40
Paris	30	6	41
London	36	4	40
Edinburgh	40	3	40

It is clear that genes A and B are both useful in distinguishing between the populations but that gene C is of little use since it shows almost no variation. Let us focus our attention for the moment only on genes A and B. By using suitable formulas, we could determine the respective values of a (1.5) and b (-4). The population living in Rome could now be represented by a value of X as follows: $X = 1.5 \times 29 + (-4) \times 8 = 11.5$. Similar values could

PRINCIPAL COMPONENTS ANALYSIS

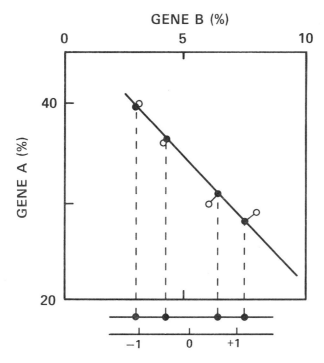

FIGURE 6.9. Illustration of a principal component. The frequencies of genes A and B for each of the four cities mentioned in the text are plotted as open circles on the graph. A straight line is drawn so as to minimize distances from these four points to the line. The original observations are replaced by their projections—the solid circles—on the line. As seen below, where the scale of the line representing the first principal component is arbitrary, the original data are now summarized in terms of a single dimension.

be calculated for Parisians ($X = 21$), Londoners ($X = 38$), and Scots ($X = 48$). Without going into a more technical discussion, a graphic representation of what is being done here is provided in figure 6.9.[8]

Moreover, we can continue this process. Having determined X, we can compute new coefficients a', b', etc. for a second index, X':

$$X' = a'x_A + b'x_B + \ldots$$

which is designed to recover as much as possible of the information that may have been lost when we computed X. If we have n genes, we can proceed until we obtain as many new index variables X, X', X'', and so forth as the original ones, progressively recovering in this way all of the information that is present in the original data.

The new variables X, X', X'' are respectively the first, second, and third principal *components*. As much information as possible is concentrated in the first principal component, as much as possible of the residual in the second, and so forth.

If we now drop the original gene frequencies and use only the X values to represent our populations, we have synthetic representations of the gene frequencies, which concentrate most of the original data in terms of a much smaller number of new variables. For example, the first component, X, may summarize 20% or perhaps even as much as 40% of the information contained in the original 39 variables. As mentioned above, part of the information that is lost can be recovered by a second component, X', and in turn by a third one, X''. The overall result is that a few new dimensions are now able to summarize the bulk of the original information. Moreover, principal components are statistically independent from one another (i.e., they are uncorrelated), which means that the information contained in each is not predictable on the basis of the other components.

This statistical technique has one severe practical limitation, however. Its application requires values for all of the genes at each of the different locations. Genes that have been known for a long time and are clinically important, such as ABO and Rh, have been studied for a large number of populations, but others are known only for a small number of populations. Not only do data have to be available for all genes and at all locations to be employed in computations, but the number of places included in the analysis must also be sufficiently large and the coverage of Europe reasonably comprehensive in order to draw contour maps of the principal components.

Filling in gaps on the map by testing people at locations without data would involve an effort and cost totally out of proportion to available research resources. A practical solution is to interpolate values for a specific location when they are unknown, using data from neighboring localities. This procedure required the construction of geographic maps for each gene similar to those seen in figures 6.4–6.6. Since drawing such maps by hand often entails arbitrary choices, using a computer is not only less subjective but also more economical. Various computer programs have been employed for this purpose.[9] Methods have also been devised to evaluate the different techniques of map construction employed. In the end, after almost two years of intensive work in collaboration with P. Menozzi and A. Piazza, maps of the first three principal

PRINCIPAL COMPONENTS MAPS

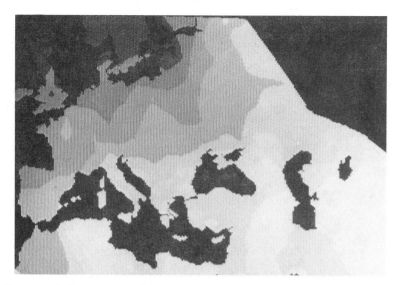

FIGURE 6.10. Contour map of the first principal component of the genetic analysis of populations in Europe. It is based upon the gene frequencies from 39 independent alleles at the human loci: ABO, Rh, MNS, Le, Fy, Hp, PGM, HLA-A, and HLA-B. Shades indicate different intensities of the first principal component, which accounts for 27% of the total variation (Menozzi et al. 1978: fig. 1).

components were obtained.[10] In the original publication of these maps (see figures 6.10–6.12) in 1978, the three components were also superimposed over one another to produce a simple composite pattern. This was done by printing each in a different color (green, blue, and red) so as to take advantage of the power of the human eye to sum elementary colors and thus synthesize the information contained in the first three components, which together account for almost 60% of the information present in the original set of genetic data. It was heartening to find that the patterns obtained were in basic agreement with those expected under the demic hypothesis.

6.7 THE PRINCIPAL COMPONENTS MAPS

The analysis shows that the first principal component collects almost 30% of the total information contained in the 39 genes. The contour map for the first principal component (see figure 6.10) presents a series of more or less regular bands running as arcs across Europe. The overall pattern of the map corresponds closely with

FIGURE 6.11. Contour map of the second principal component of the genetic analysis of populations in Europe. It is based upon the gene frequencies from 39 independent alleles as in figure 6.10. This principal component accounts for 18% of the total variation (Menozzi et al. 1978: fig. 2).

that seen for the spread of early farming. It should be stated that there is no necessary reason to link a particular principal component with a particular explanation. The fact that we are looking for precisely the pattern found in the first principal component could, of course, be a coincidence, and other explanations can never be entirely excluded. But the similarity between the two maps is so strong—a measure of the association can be obtained by computing the correlation coefficient between the values of the first principal component and the dates of arrival of early agriculture, which turns out to be quite high ($r = 0.89$)—that there is a very good chance that the two maps are functionally related.

Further support for this interpretation is offered by the results obtained when the set of 39 genes was split into two groups—one with the 21 forms of HLA genes and the other with the remaining 18 non-HLA genes—and the analysis was carried out separately for each group. The maps for the two groups are similar to one another as well as to the overall map of the first principal component based upon the full set of 39 genes. The HLA data have

PRINCIPAL COMPONENTS MAPS

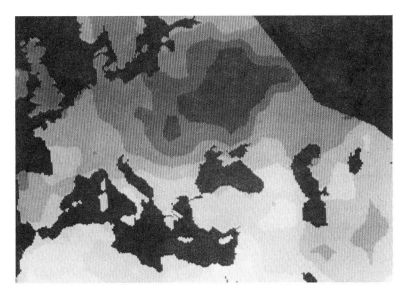

FIGURE 6.12. Contour map of the third principal component of the genetic analysis of populations in Europe. It is based upon the gene frequencies from 39 independent alleles as in figure 6.10. This principal component accounts for 11% of the total variation (Menozzi et al. 1978: fig. 3).

also been reanalyzed by an entirely different method, which is a development of a classical technique of time series analysis extended to two dimensions, with the results again agreeing with those of the previous analyses.[11] As information on more genes becomes available, there will be further opportunity to evaluate the interpretation advanced here.

The maps (figures 6.11 and 6.12) obtained for the second and third principal components, which respectively account for 18% and 12% of the variation, present patterns that can perhaps also be interpreted in terms of population movements other than the spread of early farming. The map for the second principal component shows an east-to-west gradient that corresponds to one seen for individual genes such as the B form in the ABO system and HLA-B1. A possible interpretation here would involve movement from central Asia or parts of the Soviet Union toward Europe. A whole series of migrations, perhaps starting with groups of pastoral nomads in the third millennium B.C. and continuing through historical times, are considered to have reached Europe from this part of the world.[12] In terms of the scale of their demographic impact (i.e., relative changes in population levels), the population movements involved here were probably not comparable to the spread of early farming. It may be the cumulative effect of many such population movements, each with its own more limited demographic impact, that is reflected in the map of the second principal component.

The third principal component shows an apparent peak in southern Poland and in the Ukraine. Two potential explanations can be suggested, if the pattern here is to be interpreted in terms of population movements. The first of these would be the expansion of Indo-European speaking people whose homeland has been placed in the region to the north of the Black Sea on the basis of linguistic considerations. Although this expansion has been archaeologically linked by some scholars with the Battle Axe culture dating to the third millennium B.C., considerable controversy still surrounds the question of the time depth of Indo-European languages and the means by which they became established in various parts of Europe.[13] The other explanation would be the so-called barbarian invasions in late Roman times and the period just after the collapse of the Roman Empire.[14] In this case the movements would probably have been quite modest in terms of their demographic impact. On the other hand, there would have been less time and chance for their potential effects on the gene pools of European populations to be diluted or canceled by ongoing forces of evolution. It is not our aim to evaluate these tentative interpretations of the maps of the second and third principal components in any detail here. Although it may be possible to detect subsequent migrations or population movements, these seem to be secondary in their contribution to the overall patterns of gene frequencies in European populations in comparison with the earlier demic events associated with the spread of early farming.

In closing this chapter, it is worth recalling that the first three principal components comprise about 60% of the information in the original data set of 39 genes. No lower-order principal components have been analyzed to date. The study of synthetic gene maps is still in an early stage of development, and much remains to be done. The simulation study presented in the latter part of the next chapter was developed as a means of exploring various aspects of the methods used in producing synthetic gene maps and interpreting the results obtained from them.

CHAPTER 7

SIMULATION STUDIES

7.1 Toward the Study of Process

In trying to work out the relationships between a model with a high degree of generality such as the wave of advance model and specific contexts, both archaeological and genetic, simulation studies can serve as a heuristic middle ground. A simulation study often draws attention to assumptions implicit yet previously unrecognized in the model and also to areas of knowledge needing further development. The increasing sophistication of computer applications has stimulated the use of simulation as a problem-solving technique. Computer simulation can substitute for mathematical theory when one is dealing with a problem or process that is too complicated to solve by explicit mathematical means. A well-defined problem can usually be solved in the latter way, but the solution is valid only for the particular set of numerical values that has been fed into a computer. To examine the effect that a change in a variable may have on a process, the program must be run again, perhaps many times. In addition to the expense involved in writing a computer program, which may be substantial in the case of processes of some complexity, the costs of computer runs are often high.

Against these drawbacks, computer simulations have the great advantage, common to all mathematical models, that the investigator is forced to define unambiguously all of the variables that are essential to a process. One often discovers in this way weak points in the treatment of variables and their relationships to one another, as well as points where one's knowledge of the numerical values that a variable takes is inadequate. In cases where values are unknown, trial values must be supplied if the simulation as a whole is to work. Thus, a simulation can be extremely instructive in highlighting structural weaknesses in a model and in pointing up observations that need to be made. Another advantage is that a simulation ordinarily requires no advanced mathematical training. By permitting a step-by-step account of how a process develops, a simulation model can often help to explain a complicated process even to a person who has little or no mathematical background.

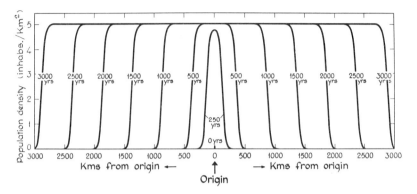

FIGURE 7.1. Computer simulation of the wave of advance employing a discrete treatment of time. The curves, which indicate the level of local population density, show the position of the wave front at 500-year intervals (Ammerman and Cavalli-Sforza 1973a: fig. 5).

We have on several occasions in the course of our study turned to simulations in order to explore questions that could not be solved otherwise. The first occasion, as mentioned in Chapter 5, arose at the very beginning of our work, when Fisher's model for the advance of an advantageous gene existed only in the form of a conjecture on the relationship between the rate of advance of a wave front and two independent variables (the rates of migratory activity and growth of a population). Preliminary estimates, made by means of a rough numerical analysis, indicated that the orders of magnitude were correct. This analysis was perhaps too simple to be called a simulation; rather, it was an approximate numerical solution of Fisher's differential equation (see figure 7.1). There was some difference between the numerical values actually obtained and Fisher's expectations. But no serious attempt was made to test whether this difference was owing to the approximation of the procedure employed or to other reasons. The results were sufficient to give confidence that the approximation due to empirical error was likely to be larger than that due to possible error in Fisher's conjecture. Moreover, it was uncertain at the time whether the solution would carry over in two dimensions, since Fisher's equation was proposed for a one-dimensional habitat. The results shown in figure 7.1 confirmed that the main expectation—that is, the formulation of a wave front advancing at a constant rate—is in practice seen in two dimensions as well. The rate of advance may be slower in the two-dimensional case by a factor of two. We have chosen to

TOWARD THE STUDY OF PROCESS

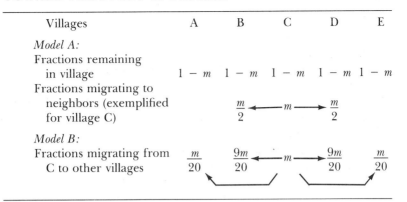

FIGURE 7.2. Two models of movement between settlements. In this abstract representation, the settlements or villages occur as points along a linear habitat. Attention is focused on village C. The proportions of individuals remaining at a village $(1 - m)$ and of those migrating to neighboring villages (m) are those mentioned in the text. In population genetics, model A is commonly referred to as a steppingstone model. Model B permits movement beyond the nearest neighbor.

ignore this source of approximation at this stage, since it is likely to be less important than others discussed in Chapter 5.

The same simulation was employed to explore other potential problems—for example, deviation from a normal (i.e., Gaussian or "Brownian") type of motion for describing migratory activity. Migration can be simulated numerically in the finite analogue of Fisher's model by making, at every generation, a fraction $(1 - m)$ of the population at a given location remain where it is, while the fraction m moves to nearest neighbor locations. In a linear habitat there would be two such locations for each population, and in a symmetric migratory scheme each of them would receive one-half of the migrants, or $m/2$ (see figure 7.2: Model A). In a two-dimensional habitat with a regular lattice there would be four such locations, each receiving $m/4$. If the units of time and space are varied appropriately, this simple scheme will generate a brownian type of migratory activity. The mean square distance of migrants will be exactly m in this case. This approach to simulating the migration component can be extended to more than one step by allowing .05 of the population, for example, to move to a second nearest neighbor location (see figure 7.2: Model B). This variable would generate a migration curve having a shape closer to the ones actually observed in human populations. As shown by the simulation, the rate of advance would then be higher than that predicted by Fisher's model, although the main property of the model—the constant rate of advance—would be retained.

Although we might expect some of these aspects of the treatment of the model to yield to mathematical solutions in the near future, many others are unlikely to find analytical solution, especially those that involve the operation of chance. As a simple example, when we move genes from one population to another, the treatment of migration presented above predicts that a fraction of a gene should move from one population to the next. In actual populations either the gene (or, more exactly, the individual carrying it) will move or it will not. The fraction of a gene (e.g., 0.7) may then be taken to mean that there is a 0.7 chance that a gene will move and 0.3 chance that it will not. In a simulation this procedure is implemented by the use of random numbers generated by the computer, which can simulate classical chance events such as the tossing of coins and the rolling of dice. Using random numbers, one can make the gene move or not with a specified probability. Mathematically, processes in which events happen on the basis of probability are called stochastic. The likelihood of finding an analytical solution for such a process is more limited than that for ordinary nonstochastic processes, which are often called deterministic. Again, computers can help in solving problems of this kind, but the time and expense required may increase by a substantial factor. Some investigators reserve the term "computer simulation" solely for models that involve stochastic processes.

By means of computer simulation, we have explored two major problems, both having to do with stochastic processes. One involves studying the process of growth and spread in the context of settlement patterns. The abstract formulation of Fisher's model becomes apparent if we consider that it either (1) treats every individual as infinitely small and assumes a uniform population density at any point (the differential equation approach used by Fisher) or (2) clusters together individuals in relatively large settlements that are considered to be the same in population size and to be located at the same distance from one another in a regular lattice. This treatment is called a "finite difference equation." When settlements are made small enough and spaced closely enough together, the second approach will give results very similar to the first. But the real situation differs from both of these treatments. The study to be described in the next section can be viewed as a *micro-simulation*. Its major aim is to see how the model works in terms of units of the process that can be observed in the archaeological record.

SIMULATION OF SETTLEMENT PATTERNS

At the other end of the scale, we wanted to test methods such as principal components analysis that can be employed for detecting a demic expansion along genetic lines. These tests entail *macro-simulation* in the sense that the basic unit is not a settlement or village but the population living in a relatively large area. In the simulation study to be described in a later part of this chapter the map of Europe will be divided into a series of cells, each covering an area of 25,600 square kilometers and each containing its own population.[1] The major bottleneck here is the space in the computer required for storing information on a large number of populations. The time and cost involved in running the simulation program depend on the total number of cells or populations on the map of Europe. In the broadest terms, the aim of the macro-simulation is to see how patterns of gene frequencies behave over space and time.

7.2 SIMULATION STUDY OF SETTLEMENT PATTERNS

The purpose of this study, which was started in the early 1970s, was to see how the abstract formulation of the wave of advance model would translate into processes of growth and spread taking place within the context of an actual settlement system. We wanted to know how the model might operate in terms of the occupation, growth, and relocation of settlements on the ground. Specifically, the study involves the simulation of the spread of the Bandkeramik culture in central Europe (described in Chapter 3). The BANDK 2 simulation model, as it is called, of a shifting form of site occupation related to the practice of swidden agriculture is based upon the interpretation by B. Soudsky and I. Pavlu, who conducted excavations at Bandkeramik sites near Bylany in Czechoslovakia.[2] Only a brief account of the main features of the simulation model and examples of the output from the simulation runs (see figures 7.3 and 7.4) will be presented here.

The settlement pattern is considered to be characterized by a linear network of sites (as in the case of sites situated along a system of linked streams with short side branches occurring at regular intervals). The site locations along the network conform to one of three states: (a) the location is occupied by a family or small village; (b) it is unoccupied and available for settlement; or (c) it is unoccupied and unavailable for settlement (i.e., the location was recently occupied and is now in a fallow state). At the start of a simulation run several locations are occupied at one end of the network, and

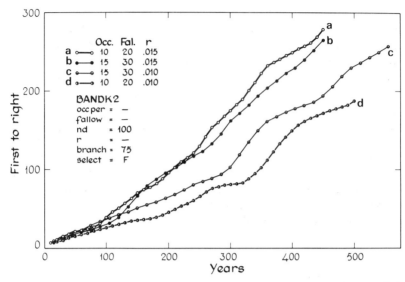

FIGURE 7.3. Examples of BANDK 2 simulation runs. The position of the occupied site furthest from the starting point of the linear network (i.e., first to right) is shown for different times in the respective runs. The variables that are allowed to change in the four runs presented here are the occupation period (Occ.), the fallow period (Fal.), and the growth rate (r). It is worth noting that the occupation period has the same ratio to the fallow period (1:2) in all four cases.

the rest of the locations are available for occupation. The individual sites experience population growth according to a logistic treatment that is related to local levels of population density. The splitting off of a new site takes place whenever it has grown to a certain size. A site location can be occupied for only a certain number of years and then passes through a fallow period of specified length. When the period of occupation comes to an end, relocation to a nearby available site location takes place.

Several examples of simulation runs using the BANDK 2 model are shown in figure 7.3, where the position of the occupied settlement furthest from the starting end of the network is plotted against different times in the respective runs. Two main variables—the growth rate and the length of the occupation-fallow period—are allowed to vary in the four runs. It is evident that a faster rate of advance (i.e., rise in the curve) occurs for the two runs based on the larger value for the growth rate, as we would expect according to the wave of advance model.

SIMULATION OF SETTLEMENT PATTERNS

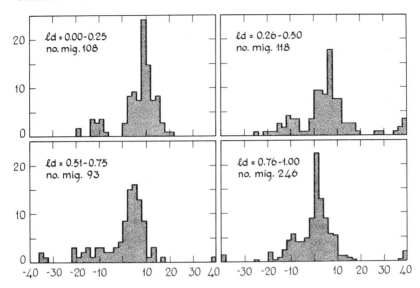

FIGURE 7.4 Migration distributions from simulation run C in figure 7.3. The frequency distribution was accumulated over the period from time 316 through 451 during the run. Distance is measured in terms of the number of site locations along the linear network. The distributions are given in terms of relative frequencies for each of four levels of local density (ld).

In figure 7.4 migration distributions obtained during the course of a run (C in figure 7.3, based on an occupation period of 15 years and a growth rate of .010%) are shown for four different levels of local density. Distance is measured in terms of the number of site locations along the linear network that a village is required to move in making a relocation (with negative distances being in the direction of the starting point and positive ones in the direction of the advance). Local density (ld) is measured in terms of the relative proportion of locations within a specific neighborhood of a given site location that are currently occupied or in fallow. The pioneer zone, where population growth is taking place most actively, would correspond to the area with a local density falling in the range between .00 and .25. It is interesting to note the progressive shifts in the migration distributions to shorter distances with an increasing level of local density. In the pioneer zone the mode of the "positive" relocations is 9-10 site locations long. In those areas with high local densities (i.e., 0.76–1.00, where, incidentally, there is relatively little population growth), it is only 2-3 site locations long. One of the striking features of the four distributions taken as a whole is that only a relatively small proportion of the relocations is larger than 6 site locations in either direction.

Without going into the question of evaluating the BANDK 2 model and the need for exploring alternative models, it would seem to be clear that micro-simulation provides insight into how the wave of advance model may operate within the context of a settlement system.[3] There appears to be no loss of the main properties that enter into the formulation of the mathematical model as we move closer toward what can be observed in the archaeological record. One point shown by the simulation is that growth need be taking place actively only in frontier areas where the spread is actually occurring. In fact, if growth does not slow down in those areas behind the advancing front, it will lead either to overcrowding over the long run or to the relocation of sites over unreasonably long distances. Another point illustrated by the simulation is that only modest levels of migratory activity are indeed required to sustain a demic event such as the spread of early farming.

7.3 Modeling Population Interactions

One aspect of the spread of early farming that requires further attention is the interaction between hunter-gatherer and farming populations. Some discussion of this question as it relates to archaeological evidence was presented in Chapter 3. In this section we shall develop a more abstract treatment of population interactions in preparation for the macro-simulation study to be discussed in the next section. Readers familiar with Lotka-Volterra equations, which are used for studying interactions between species in ecology, will recognize the formal treatment that we have adopted.[4]

Population growth is considered to take place separately for farmers and hunter-gatherers according to logistic models where the parameters defining the initial growth rate and the density level at saturation may differ for the two populations. Specifically, the density level would be higher among farming populations.

	Hunters	*Farmers*
Initial growth rate	α_H	α_F
Number or density at saturation	H_{max}	F_{max}

Assuming that there is a sufficiently large area, the two populations can live side by side, and their coexistence will be favored if they occupy somewhat different ecological niches. A situation of this

MODELING POPULATION INTERACTIONS

kind is still found in a few ethnographic situations, such as that of pygmies and Bantu farmers in parts of Central Africa. Under the standard logistic equation, the growth rate of the two populations can be represented as:

for hunter-gatherers
$$\frac{dH}{dt} = \alpha_H H(1 - H/Hmax)$$
for farmers
$$\frac{dF}{dt} = \alpha_F F(1 - F/Fmax)$$

The two populations, although essentially living and reproducing independently in their own subareas, inevitably interact. In agreement with the simplest assumption commonly used in such models, interaction terms will be considered to be proportional to the product of the number of farmers F and the number of hunters H and a coefficient of proportionality B that measures the interaction of the two populations. It is useful to identify four possible forms that interaction can take.

Acculturation. This process involves the transition from one type of economy and set of customs to another—in the present case, hunter-gatherers becoming farmers. In terms of population dynamics, this transition will involve the addition of a number of $B_A FH$ to the farmers' rate of change and the subtraction of the same term from the hunters' rate of change. The rate of *acculturation*, B_A, is positive.

Warfare. This factor will decrease the rates of change of farmers and hunters by $B_{WF}FH$ and $B_{WH}FH$ respectively, where B_{WF} and B_{WH} are the rates of loss of the two sides due to warfare. This formulation, however, presupposes a condition of perpetual warfare, which is unlikely. A more satisfactory formulation would be one in which the B_W's are not constant but decrease over time.

Disease. This factor will also decrease both rates by amounts dependent on B_{DF} and B_{DH}. It is known that small, isolated communities are often vulnerable to epidemic diseases determined by contact with larger populations. Other sources of disease can operate in both directions.

Mutualism. Trade and other positive interactions increase the rates of change of the two populations; B_{MF} and B_{MH} are positive quantities measuring the additions to the respective rates of change of H's and F's, which derive from mutualistic exchanges.

Altogether, the rates of change of the population of farmers F

and that of hunter-gatherers H will be modified by these interactions as follows:

$$\frac{dH}{dt} = \alpha_H H(1-H/H_{max}) - B_A HF - B_{WH} HF - B_{DH} HF + B_{MH} HF$$

$$\frac{dF}{dt} = \alpha_F F(1-F/F_{max}) + B_A HF - B_{WF} HF - B_{DF} HA + B_{MF} HF$$

These are the Lotka-Volterra equations developed for the study of population interactions. They can be put in the usual form:

$$\frac{dH}{dt} = \alpha_H H(1-H/H_{max}) + \beta_H HF$$

$$\frac{dF}{dt} = \alpha_F F(1 - F/F_{max}) + \beta_F HF$$

where

$$\beta_H = -B_A - B_{WH} - B_{DH} + B_{MH}$$
$$\beta_F = B_A - B_{WF} - B_{DF} + B_{MF}$$

In the general case, the final outcome of the interaction between the two populations depends on the values of the B's and the saturation values. The α's are unimportant if they do not differ from one another. In some cases the fate of the interaction may depend also on the initial values of the two populations. For the present case, the analysis will be considerably simplified, since the effects of mutualism, warfare, and disease are probably small compared with the acculturation rate, so that β_H is dominated by acculturation and hence likely to be negative, while β_F is positive. This model will usually leave only few or no hunters at equilibrium. Small populations easily disappear under chance fluctuation. Therefore, the usual outcome expected under the model is the survival of farmers and the extinction of hunters or, rather, their absorption into the farming community. The most likely situation favoring survival of hunters side by side with farmers under such conditions would be one in which the two populations occupy only partially overlapping niches. It is worth adding that such a case is not envisaged in the classical Lotka-Volterra equations and would require special treatment.

In the macro-simulation study described in the next section, we shall simplify the model of interaction by assuming that the only component is that of acculturation, B_A. Moreover, in the treatment adopted, acculturation (i.e, the shift to food production by hunters and gatherers) and gene flow (i.e., the addition of hunter-gatherer genes to the gene pool of farmers) occur simultaneously in a single step. In other words, when hunters and gatherers experience acculturation, they immediately join the population of farmers. In the real situation, the two steps may take place separately with some time between them. This simplification, which is convenient for studying the basic development of genetic patterns, means that one constant can be used to express jointly both the rate of acculturation and gene flow.

7.4 Simulation Study of Genetic Patterns

The motivation for attempting the macro-simulation was to explore the importance of different variables in generating and maintaining genetic patterns related to major demic events and to evaluate the power of methods such as principal components analysis for detecting demic clines. The simulation model that we have used will be described in some detail later in this section. The basic unit of the macro-simulation is a geographic area or cell on a map that measures 150 kilometers on a side and contains a population whose gene frequencies are traced over time. In an initial version of the simulation, which was developed in the early 1970s, the map contained a total of 121 cells arranged in a square lattice (11 x 11).[5] In a more recent version, 816 cells are arranged in a rectangular lattice (34 x 24). The latter covers a total surface area that is not far from that of Europe. About half of the cells in this case, corresponding roughly to marine areas in a rather stylized but still reasonably realistic representation of Europe, are actually left unoccupied, so that only about 400 units are treated as having populations. In each of the cells there is at the beginning only a population of hunters and gatherers (H), although room is left open for a population of farmers (F) who will arrive at a later point in time. For each spatial unit, the quantities that are subject to change during the course of a simulation run are the number of individuals alive at a given time (which are kept separately for populations of H and F) and the gene frequencies of ten (or more) genes, each of which can exist in two or more different forms (alleles). For instance, for one gene that occurs in three different forms (i.e., a gene similar to ABO), gene frequencies are set initially to a value of 33.3% for each form. The time cycle employed in the simulation is one generation, which is taken to last 25 years.

During the course of a simulation run, four phases of "evolution" can be distinguished. During the first, the populations of hunters and gatherers are allowed to differentiate under genetic drift. The second phase sees the initial appearance of food production in southwest Asia, with farming populations being allowed to grow and expand geographically. In the third phase, which begins after the completion of the spread of early farming to Europe, subsequent population radiations are set in motion from other places on the map. The aim here is to see whether such events obliterate the patterns associated with the preceding expansion of farming and the extent to which they can be distinguished. During the fourth phase, time is allowed to elapse without further major demic events occurring in order to see the effects of the passage of time on genetic patterns. This phase is a check, since our observations on the gene patterns are based on living populations and thus on data representing a point in time several thousand years after the events of interest. Following a basic account of each of the four phases, a brief sketch will describe the routines employed for operating the processes of growth, migration, and acculturation during each cycle of the simulation.

Phase I. 11,000–8000 B.C.:
Genetic differentiation among populations of
hunters and gatherers

In this phase, a population of hunters and gatherers occupies each of the cells on the map. At the beginning, a population of 500 people is assigned to each spatial unit.[6] This figure is considered to be the average size of a population of hunters and gatherers; the density of each spatial unit would be .02 people/km^2. During this phase, evolution is based on a model of differentiation between areas due to random genetic drift, which is partially buffered by the allowance of some migration between neighboring populations.[7] All 400 cells initially have the same gene frequencies. With the passage of time, however, populations begin to diverge, and differentiation is allowed to proceed for as many generations as is necessary in order to obtain a degree of genetic variation between some populations comparable to that observed today between two neighboring continents (such as Europe and southwestern Asia). The rate at which genetic differentiation proceeds is inversely proportional to the size of populations and also to the migration rate between neighboring populations, which was set to $m = 4\%$ per generation.[8] Under these conditions, it takes some 120 to 150 generations (i.e., some 3,000 years) for the variation between gene frequencies to rise to the desired level. An example of the geographic pattern of a gene frequency map generated in this way is shown in figure 7.5. On some parts of the map, high frequencies are observed; on other parts, low ones. In general, neighboring populations have values that are similar, but occasionally even adjacent populations have quite different gene frequencies. No very clear pattern emerges from this picture, other than the effects of chance in varying local gene frequencies.

Phase II. 8000–4000 B.C.:
Spread of early farming from the Near East

Once the variation between populations of hunters and gatherers has reached the desired level, the transition to food production is set in motion. This event is taken to occur at a date of about 10,000 years ago. A few neighboring cells in southwest Asia are chosen as the place where agriculture begins, and their populations are now considered to be farmers. The growth parameters are changed accordingly from those of hunters and gatherers to those of farmers (i.e., the growth rate of $\alpha_H = 0.25$ is changed to $\alpha_F = 0.50$ per generation, corresponding to a change from 0.9% to 1.6% per year, and the saturation level increased from $H_{max} = 500$ to $F_{max} = 10,000$ or a density of 0.4 people/km²). Hunters and gatherers are considered to be totally absent from those cells where farming originates; in other words, the gene frequencies of previous hunters and gatherers now become those of farmers. In all other cells on the map populations of hunters and gatherers keep their size, growth parameters, and gene frequencies. But room is now made available in each cell for the arrival of farmers, who live as a separate population alongside the local population of hunters and gatherers. As a consequence of population growth and migratory activity, farmers expand into new cells. With the arrival of farmers in a cell, there arises the possibility of the acculturation of local hunters and gatherers to the new way of life according to the model described in the previous section. Within a cell, acculturation increases as the farming population grows in size, in agreement with the model of hunter-gatherer acculturation described above.

Since movement in the simulation is allowed to happen only between adjacent cells, it takes a good deal of time before farming populations can reach cells in northwest Europe (see figure 7.6). The most distant cell on the map is located some 38 steps away from the nuclear zone in the Near East. In fact, it takes many more than 38 generations in the simulation of this phase (about three times as many, or some 3,000 years) for the expansion to be com-

							64	47	69	40			44	56	39	38	25	33	29	34	35	56	33	31	75	56	65	00	44	85	74	49	16	07
							33	94	68	97			22	59	43	52	78	86	77	38	36	55	52	63	52	50	41	31	52	74	69	30	57	03
		23	37							36			37	70	41	93	52	89	53	42	34	43	44	36	56	83	67	63	22	45	35	52	43	05
		29	33					35	52	54			76	66	45	43	24	65	60	24	35	28	85	84	81	67	68	58				51	40	12
	59	46	38	43				59	62		50	78	51	54	62	43	56	52	75	60	43	84	75	97	45	64	56	38					19	M
	62		39	63			38	27	22	16	25	50	51	49	71	63	84	71	66	89	60		M	34	24									..
	48		40	50	26		48	43	43	33	46	22	16	39	88	M	54	52	41	40	37	15	00		46	27	36	41	37	69				..
			51	97	64	82	50	31	36	60	39	28			66			68	39		05					81	38	40	39	83	65	84	M	00
					51	54	64	56	28	39	41	96	64	07	00	..		38	86										87	46	92	12	28	06
			00	50	27	42	61	34			60	95	74	27			61	51	50			27	44	44	63	47	42	24	63	65	38	22	23	29
				45	33	49	81			83	46	78	55	33	36	47	67	48	37		49	39	67	76	48	49	60	58	52	33	45	41	32	59
				03	56	47	37	11	00	00	09		47	56	44	91	96	35	35	01	33	86	36	36	35	60	56	52	41	65	37	46	47	41
				27	47	21	59	25			00	16			67	50	49			00	15	88	59	53		45	50	34	46	65	29	36	37	72
95	41	49	52	44		33						24	47	47	50	30	48	32		..	00	M				28	50	55	43	48	50	50	39	31
69	40	61	35	35	26									12			33	54								35	43	31	50	20	65	60	83	81
55	49	40	46	34								01	41	45												34	38	31	24	33	24	54	77	26
36	34	58	60	57							89	71	54												29	39	31	35	25	56	42	60	48	32
23	70	28	25					52	04	15	80												62	47	59	44	23	58	70	64	90	26	21	68
		35			39	33	88	68	36	45	58							85	24	19	26	18	40	37	25		97	77	67	93	18	38	37	56
	63	94	61	25	16	68	78	52	39	36	44	53	41					52	48	37	37	39	21	31	25			58	99	99	28	28	56	69
60	55	84	44	38	09	13	77	49	57	33	29	33	23	31	94	65	97	62	44	72	56	65	66	53	61	52				85	45	35	66	66
51	31	46	36	60	55	19	36	50	43	50	56	42	55	61	89	M	90	35	39	09	43	65	68	87	43	38	45			..	26	63	57	71
93	43	55	42	34	33	47	23	36	76	43	31	60	68	38	21	66	85	40	83	36	72	72	46	51	53	63	34	13		00	16	49	42	39
66	48	49	45	23	22	54	74	16	41	48	34	34	66	58	45	45	84	55	46	76	90	M	26	25	40	51	43	37		00	27	18	28	75

FIGURE 7.5. Example of the simulation of genetic patterns in pre-neolithic populations in Europe. During Phase I of the macro-simulation described in the text, the force of evolution that is operating is considered to be random genetic drift. The macro-simulation of Europe here represented visualizes a 35 x 24 lattice of hunter-gatherer populations. Seas and mountains are left blank. Each two-digit number represents the gene frequency in percent in the local population of one gene undergoing random genetic drift (with some migration to nearest neighbors). *M* indicates 100% gene frequency and . . indicates extinction of the local population. Considerable variation in the values of gene frequencies can be observed from one part of the map to the next (courtesy of Rendine et al. 1984).

pleted. There has to be sufficient time for the growth of the farming population within a newly started cell before any substantial number of farmers can immigrate to an adjacent cell. The migration rate of the farming population to neighboring cells is set at the same value as that for hunter-gatherers ($m = .04$), which is drawn from ethnographic evidence discussed in Chapter 5. This value interestingly gives a rate of advance that corresponds closely with that observed for the spread of early farming in Europe. Figure 7.6 shows the expansion at four points in time in this simulation of the neolithic transition. An example of the geographic distribution of one gene frequency after the expansion is given in figure 7.7: at the end of Phase II it showed a substantial difference in frequency between populations in southwest Asia, where values were low, and those in northwestern Europe, where (perhaps not unlike the case of the actual Rh − gene) values were high.

Phase III. 4000–0 B.C.:
Period of subsequent demic events

During this phase, subsequent demic events are considered to take place. In particular, two such events are allowed to occur after the spread of early farming. The object was to see whether the genetic pattern associated with the demic expansion of Phase II persists after subsequent population movements and can still be recognized. Without making any attempt to model these events on the basis of archaeological and historical evidence, we merely introduced two such events, located where they appeared to originate on the principal component maps presented in Chapter 6.[9] The same "model" of demic diffusion was used for these episodes, even though actual developments are likely to have taken forms quite different from those involved in the spread of early farming. It is reasonable to think that the shifts in levels of population density associated with later demic events were in general of a much smaller order than those associated with the neolithic transition, including the ongoing growth among local neolithic populations up to the time when subsequent (hypothesized) demic episodes begin to occur.

Phase IV. 0–2000 A.D.:
Ongoing evolution to the present time

During this phase no other major migration is considered to occur, but reciprocal migration between neighboring cells continues to

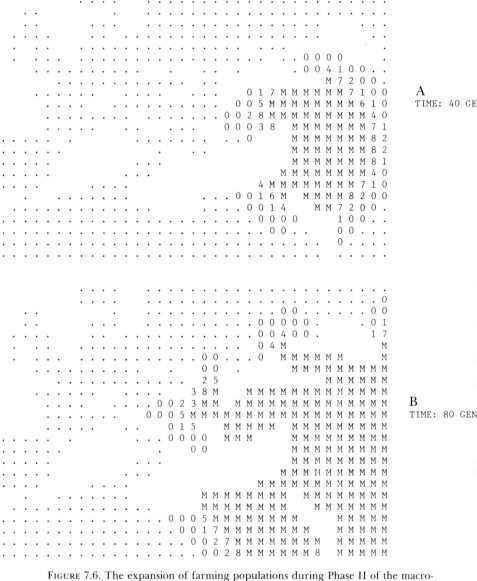

FIGURE 7.6. The expansion of farming populations during Phase II of the macro-simulation. The maps show the spread at four points in time as measured in terms of generations (a generation of 25 years is taken to be the cycle of the simulation): A = 40 generations (1,000 yrs.) after the beginning of the spread; B = 80 generations (2,000 yrs.); C = 120 generations (3,000 yrs.); D = 160 gen-

```
. . . .              . . . . 0 0 0 2 7 M M M M M M 6 7 M M M M
  . . . .          . . 0 0 0 1 3 7 M M M M M M M M M M M M M
. .         .       . 0 0 0 4 8 M M M M M M M M M M M M M M M
. .       . . .       0 0 1 5 M M M M M M M M M M M             M M M
. . . .         . 0 0 0 5 M M M M M M M M M M M                 M M
. . . .       . . . 0 0 0 3 M M M M M M       M M M                 M
. . . . .     . . . 0 0 1 1 7 M M M M M M M M     M M M M M           M
. . . . . .   . 0 0 1 4       M     M M     M             M M M M M M M M
. . . .   . 0 0 6 8 M M M 7     M M                 M M M M M M M
. . . . .     M M M M       M M M         M M M M M M M M M M M M
. . . .       8 M M M M M M M M       M M M M M M M M M M M M M M      C
. . . . 0 0 4 8       M M M M M M M M M M M M M M M M M M M M M    TIME: 120 GEN.
. . . .           3 8         M M M       M M M M     M M M M M M M M M
                    M M M M M M M       M M M           M M M M M M M M
. . . . .                 M         M M               M M M M M M M M
. . . . .             1 7 M                         M M M M M M M M
. . . .             0 0 1                         M M M M M M M M M
. . .             . . . .                       M M M M M M M M M
. . .         . . . . 0 0                 M M M M M M M     M M M M M M M
. . . . . . . . 0 0 1 6 M             M M M M M M M M         M M M M M M
. . . . . . . . 0 0 3 M M M M M M M M M M M M M M               M M M M M
. . . . . . . . 0 0 2 8 M M M M M M M M M M M M M M M             M M M M M
. . . . . . . . 0 0 0 5 M M M M M M M M M M M M M M M M           M M M M M
. . . . . . . . 0 0 1 7 M M M M M M M M M M M M M M M M             M M M M M


              . . . 0         M M M M M M M M M M M M M M M M M M M M M
              . . . 0         M M M M M M M M M M M M M M M M M M M M M
    . .       . . . 0         M M M M M M M M M M M M M M M M M M M M
    . .           M M 6       M M M M M M M M M M M M M M       M M M
  . . . .           M M       M M M M M M M M M M M M M M M       M M
.   0 0         M M M M M M M M M M M M M M     M M M                 M
.   0 0 2     M M M M M M M M M M M M M M M       M M M M M             M
  0 2 9 M M M M M M M         M       M M M         M                 M M M M M
      M M M M M M M M M M M M   M M                         M M M M M M
    0 2 8 M M M         M M M M       M M M         M M M M M M M M M M M
    5 M M M         M M M M M M M M         M M M M M M M M M M M M M       D
    8 M M M M M M M     M M M M M M M M M M M M M M M M M M M M M    TIME: 160 GEN.
    6 M M M M         M M       M M M         M M M M     M M M M M M M M M
.   0 0 0       M             M M M M M M       M M M             M M M M M M M M
. .   0 0 0                         M             M M                 M M M M M M M M
. . . .                       M M M                         M M M M M M M
. . .                         M M M                         M M M M M M M M M
.               M M M M                         M M M M M M M M M M M
.     4 M M M M M                     M M M M M M M       M M M M M M M
. 0 0 1 6 M M M M M M M M                 M M M M M M M M       M M M M M M
. . 0 0 2 8 M M M M M M M M M M M M M M M M M M M M M             M M M M M
. . 0 0 1 8 M M M M M M M M M M M M M M M M M M M M M M             M M M M M
. . 0 0 0 6 M M M M M M M M M M M M M M M M M M M M M M M             M M M M M
. . 0 0 0 2 M M M M M M M M M M M M M M M M M M M M M M             M M M M M
```

erations (4,000 yrs.). A dot indicates that there are no farmers in that location; M indicates that the saturation level (10,000 individuals) has been reached. Numbers 0-8 indicate 0-10%, 10-20%, ..., 80-90% of the saturation level have been reached (courtesy of Rendine et al. 1984).

take place as it did in the three previous periods. Under its action, the gradients of gene frequencies established by the demic episodes introduced during Phases II and III are expected to be partially eroded. The aim of this part of the experiment is to observe the effects of the passage of time on the patterns of gene maps.

It is useful for purposes of clarity to describe briefly the sequence of events taking place during each cycle (generation) of the simulation. The following events happen to one or both of the populations living in a given cell.

1. *Growth*. Population growth occurs according to the appropriate logistic model, with different parameters for hunter-gatherers (H) and farmers (F). The hunter-gatherers are already at saturation at the start of the first phase and stay at this level until acculturation eventually transforms them into farmers. The population of farmers increases until saturation is reached. The earliest farmers, as mentioned above, are the original hunter-gatherers in selected cells in the area representing the Near East on the map. Every growth cycle involves the production of a new generation of people. Thus the gene pool of each population living in a cell is completely renewed. The frequency of each gene in a population is replaced by a new and usually slightly different one due mainly to random genetic drift.[10]

2. *Migration*. A fraction m of each population living in a given cell is allowed to immigrate to neighboring cells. Individuals can move only between populations of the same kind: that is, hunter-gatherers move to neighboring populations of hunters and gatherers and likewise for farmers. The migrants leaving a cell are distributed equally among neighboring cells on the map. Genes allotted to these migrants are drawn as a random sample from those existing in the "donor" population.

3. *Acculturation*. In the simulation, only the simplest form of interaction between populations is considered to occur: namely, the transfer of H individuals to the population of farmers (F) within the same area. As we have seen in section 7.3, this happens with a probability defined by the acculturation rate per generation, B. The actual value chosen for B in the simulation can have a major effect on genetic gradients generated by the neolithic transition. This is in effect a more refined way of treating the question of the different genetic patterns that one would expect under cultural, demic, or mixed models of diffusion (discussed at the end of Chapter 5). The simplest way to measure the residual gene pool of a farming population after migration is to plot the percentage of an ideal gene

														65	68	67	45		52	58	67	60	45	46	43	51	52	53	61	62	69	68	67	58	62	57	55	62	58	55
														80	71	66	54		48	55	53	48	48	43	39	47	55	58	62	72	74	71	69	59	55	56	55	52	53	53
																	48		44	51	47	39	41	40	39	49	53	59	71	73	79	71	63	65	51	51	60	49	45	48
				73	85									67	62	65			50	51	45	41	48	39	39	49	46	56	67	74	74	69	66	62				44	41	42
				86	88										75		56		50	50	49	44		34	36	42	44	56	62	73	79	73	66	69						
			73	86	74	63									69	74	65	63	49	46	47	46	40	34	34	34	37	51	61	74									57	53
			68	63	58									72	66	63	66	67	50	43	42		35	36	33	33	31	39	61	69	63									51
			76	73	73		70							69	63	50	52	55		43	43		70	36	33	33	31	32	45	69	72									51
				69	74	63	70							59	63	52	56	56		43		70	35	32	29	32					63					43		43	41	40
						55	70							53	62	61	58	54	48	44				33	30	29					56			40	44		44	45	40	40
						52	63	66						50				55	45	42	37	40	35	35	30							38	36			42	40	43	41	40
				21	24	35	36							35	36	40	47	42	42	41	37	41	39	35		45	40	40	43	40	38	40	37	42	39	41	46	45	35	35
					19	26	39							29	24	27	36	36	43	41	41	39	40	31	30	26	33	33	40	41	43	37	44	38	36	40	39	34	31	35
					30	26	31							28	28		32					42	46	36	31	30	32	31	39	36	40	40	39	40	30	36	30	25	23	27
					31	25								29	28				25			39	42		30	32	32	35		30		38	32	38	30	34	34	29	27	30
							26							20					25			37	43	42	28	28								29	30	26	28	30	30	36
28	30	30	38	36	32															32	32	49		42						45			35	31	29	30	31	32	30	33
29	30	36	36	30	29												36		43	28			53	50						51	47	51	38	34	32	29	32	30	31	32
31	25	33	32	30	31												39		43	29									48	54	54	52	37	35	32	28	27	28	28	29
21	27	37	34	28													36		44	39									52	57	53	58	38	32	32	28	31	25	31	33
33	33	35	42													32	43	42	36					30	35	37	47	46	46	53	53		32		25	24	29	28	28	29
			49					28	30							27	33	35	32					37	36	36	46	51	52	57	52	53					29	25	31	33
				43	39	27	27	28	30							36	33	32	28	29						38	47	51	53	57	53	57		32		25	29	31	36	32
			37	33	23	18	23	24	28							26	35	35	30	30	42	49	46	44	35	36	51	55	60	58	53	55			25	25	28	29	31	36
			23	20	18	16	24	25	27							20	31	36	29		45	45	51	47	36	62	62	53	55	51	44					28	31	32	38	40
			23	21	21	27	27	23	32							31	37	37	33	44	45	54	50	54		60	57	61	60	58	50	52	48	54		29	31	33	35	39
			35	40	45	45	49	42	45							42	45	39	44	49	50	57	56	57	55	60	64	66	58	59	47	51				34	33	37	40	42

FIGURE 7.7. The geographic distribution in Europe of one of the 20 genes from the macro-simulation study. This example shows a pattern at the end of Phase II, or the time when the expansion of early farming over Europe has been completed. Symbols as in 7.5 (courtesy of Rendine et al. 1984).

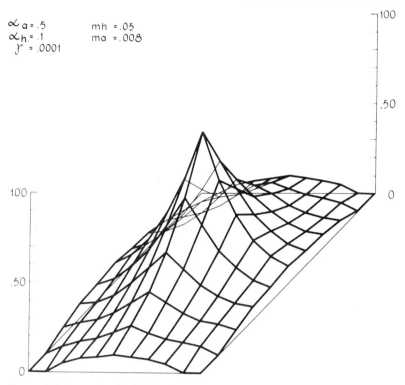

FIGURE 7.8. Example of a demic cline. The vertical axis indicates the percentage of the genes of "early farmers" at generation 55 within populations located at varying distances from the starting point of the demic expansion. In the earlier macro-simulation study by Sgaramella-Zonta and Cavalli-Sforza (1973), which is shown here, this point was taken to be the center of an 11 x 11 lattice of cells, each having its own population.

assumed to be at frequency 100% in the cells where agriculture originated and 0% in the other cells (i.e., those containing populations of hunter-gatherers). This model is illustrated by a plot from the first macro-simulation study done with L. Sgaramella-Zonta, as shown in figure 7.8. As one might expect, the highest gene frequencies are located in the center of the 11 x 11 lattice employed in the earlier study (see section 7.1), where farming is considered to start in this case. With the given set of values used in this simulation run, frequency decreases regularly until it reaches almost 0% at the periphery.

We have introduced the term "demic cline" for curves of the type shown in figure 7.9, which indicate the proportion of the initial

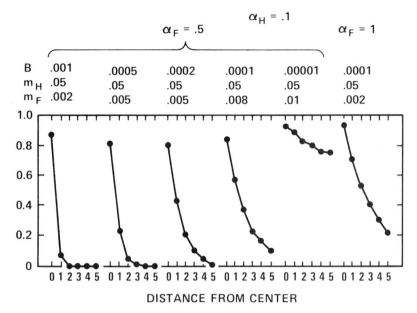

FIGURE 7.9. Demic clines under different combinations of parameters. The curves all derive from the earlier macro-simulation study by Sgaramella-Zonta and Cavalli-Sforza (1973) and are shown at generation 55 in the simulation runs. The six points forming each curve correspond to six contiguous populations in a row or column starting at the center of the (11 x 11) square lattice where the demic event was initiated. The curve representing the demic cline shown in Figure 7.8 is the fourth from the left in this figure. The value of α_F is mostly 0.5, that of α_H, 0.1. The values of the acculturation coefficient B and of the migration rates of hunters to neighboring hunters and of farmers to neighboring farmers (m_H, m_F) are as shown in the three rows above the figure. The ordinate expresses the proportion of farmers' genes at each distance from the center of spread.

farmers' genes observed in populations after the spread as one moves away from the center. In this figure, the frequency of a gene starting on the left with its value at the center is plotted with increasing distance from the center. The different demic clines presented in the figure were obtained by using other values of α_F and B in the initial simulation study. As can be seen, there is a clear response of the slope of the demic cline to the acculturation rate B. When this rate is highest ($B = .001$), few of the genes associated with initial farming populations (i.e., at the center) become incorporated in other populations on the map. In other words, most of the farming population in a given cell derives from hunters and gatherers who have learned to farm. With the lowest B value tested, which is 100 times smaller, the gene pools produced are dominated by the genes of the initial farmers—about 90% of such genes appear in the central part of the lattice and about 75% at the periphery. A gradient of gene frequencies can still be recognized in the latter case, but it is a rather flat one. Clearly, a wide range of resulting genetic patterns can be obtained between these extremes. Interestingly, the variable that appears to play the dominant role here is the rate of acculturation.

7.5 Evaluating Principal Components Analysis as a Method

Macro-simulation is instructive from a heuristic point of view in showing how genetic patterns can be generated by demic events, but its main contribution may be to provide a means for evaluating methods used, such as principal components analysis, and to help in the interpretation of the results obtained. This aspect of our study has only just begun. Without going into a detailed evaluation here, the macro-simulation does provide insight into central questions: Can principal components analysis unravel separate demic events of the type proposed for the spread of early farming? Can it help in understanding the kinetics of the process? Previously, we had optimistically assumed that it could do so and believed that a demic explanation was the most likely one for the patterns recognized when the method was applied. On the basis of the simulation work, we are now on firmer ground in giving answers to the questions raised above.

As an example, a map of the first principal component derived from an analysis of the genetic data produced by a run of the macro-simulation after Phase IV is shown in figure 7.10. It shows a series of contours that seem to be linked primarily with the demic event initiated in Phase II of the simulation. Ideally, we would like to be in a position where the results of the principal components analysis could be used to make inferences about the date of a major demic event and its main demographic characteristics. This is perhaps asking for too much. We may not be able to disentangle one from the other without turning to outside sources of information. On the basis of principal components analysis alone, we may be able only to identify their joint effect. The implication is that major patterns detected in synthetic gene maps probably have meaningful

							1	1	1	1			5	5	5	5	5	5	5	5	5	5	5	5	5	4	4	5	5	4	4	4	4	4
							1	1	1	1			5	5	5	5	5	5	5	5	5	5	5	5	5	5	5	4	4	4	4	4	4	4
		3	3	3						1			5	5	5	5	5	5	5	5	5	5	5	5	5	5	5	5	4	4	4	4	4	4
		3	3					1	1	1			4	5	5	5	5	5	5	5	5	5	5	5	5	5	5	5				4	4	4
	4	3	3	3				1	1		3	3	4	4	5	5	5	5	5	5	5	5	5	5	5	5	5	5					5	5
	4		2	2			1	1	2	2	3	3	4	4	5	5	5	5	5	5	5		5	5	5									5
	4		2	2	2		1	1	2	2	3	3	4	4	5	5	5	6	6	5	5	5	5		6	7	7	7	7	7				6
			2	2	2	1	1	2	2	2	3	3			4			6	6		5					7	7	7	7	8	8	7	7	7
					1	1	1	2	2	2	3	3	4	4	4	4		6	6										8	8	8	8	7	7
			1	1	1	1	1	1			3	3	4	4			7	7	7			9	9	9	9	9	9	8	8	8	8	8	8	8
				1	1	1	1			3	3	4	4	4	5	6	6	7	7		9	9	9	9	9	9	9	9	8	8	8	8	8	8
				1	1	1	1	2	2	3	3		5	5	5	6	6	7	7	8	9	9	9	9	9	9	9	9	9	8	8	8	8	8
				1	1	1	1	1			3	4			5	6	6			8	9	9	9	9		9	9	9	9	8	8	8	8	8
1	1	1	1	1		1						4	4	5	5	6	6	6		9	9	9				9	9	9	9	9	9	8	8	8
1	1	1	1	1	1									4			6	6								9	9	9	9	9	9	9	8	8
1	1	1	1	1								3	3	4												9	9	9	9	9	9	9	9	9
1	1	1	1	1							3	3	3												8	8	9	9	9	9	9	9	9	9
1	1	1	1					2	2	2	2												7	7	8	8	9	9	9	9	9	9	9	9
		2			3	3	3	3	2	2	2							5	5	5	6	6	7	7	7		9	9	9	9	9	9	9	9
	3	3	3	3	3	3	3	3	2	2	2	2	2					5	5	5	6	6	6	7	7			9	9	9	8	8	9	9
4	4	4	3	3	3	3	3	3	3	2	2	2	2	3	3	4	4	4	5	5	5	6	6	6	6	6				8	8	8	8	8
4	4	4	4	3	3	3	3	3	2	2	2	2	2	3	3	4	4	4	5	5	5	6	6	6	6	6	6			8	8	8	8	8
4	4	4	4	4	3	3	3	3	2	2	2	2	3	3	3	4	4	4	5	5	5	6	6	6	6	6	6	6		8	8	8	8	8
4	4	4	4	4	3	3	3	3	3	3	2	2	3	3	3	4	4	4	5	5	5	5	6	6	6	6	6	6		8	8	8	8	8

FIGURE 7.10. Example of the first principal component when an analysis of gene frequencies is conducted for populations in the macro-simulation study by S. Rendine et al. (see figures 7.5–7.7). Each of the populations has 20 genes. The analysis is done here after Phase IV (courtesy of Rendine et al. 1984).

histories behind them, but a breadth of knowledge and some patience may be required to discover what they are.

An adequate number of genes is essential to an analysis, if the principal components method is to yield useful results. We also have to be prepared for the possibility that some principal components in an analysis may reveal unsuspected sources of variation. Major patterns of natural selection might be detected when data from actual populations are analyzed. In fact, analysis of gene frequencies on a worldwide basis has made it possible to show that there is a gradient that appears to correlate closely with distance from the equator and hence climate.[11] When only a part of the world is examined, no major effect of climate is detectable, at least in terms of the first three principal components.

Principal components analysis is by no means new. It has been applied to a wide range of problems ever since computer programs for this type of analysis became widely available. But it is worth emphasizing that the so-called maps of the first two or three leading components that are in common use have no relationship to our contour maps of individual components. The latter are much more powerful in suggesting interpretations of phenomena that have influenced (usually) a large number of genes. In principle, it is reasonable to expect that a great deal about underlying sources of variation can be learned from studying the contour maps of principal components. Although some of the sources of variation may be difficult to interpret, any major principal component is likely to have behind it some historical or ecological explanation.

CHAPTER 8

CONCLUSIONS

The results of our investigation encourage us to think that a bridge can be established between subjects as seemingly diverse as archaeology and genetics. We find that it is possible to interpret patterns of the geographic distribution of genes, something that has long puzzled geneticists, on the basis of events observed in the archaeological record. At the same time, insight into developments of archaeological interest, such as the origins and spread of agriculture, can be obtained by analyzing genetic data and using conceptual frameworks drawn from the study of population biology. In this chapter we would like to review some of the main results of our collaboration, as well as draw attention to future prospects and problems. We want to emphasize that what we have presented in the preceding chapters should in many cases be regarded as research in progress. Both of the fields in which we work are undergoing rapid development. Therefore, we can give only a cross section in time of the investigations that will, we hope, continue to evolve at a substantial rate.

We begin with archaeological results and prospects, which initially stimulated the study. We have seen that it is possible to measure the rate of spread of early farming in Europe. This success was due in part to the development of simple but appropriate methods for undertaking such a rate measurement and in part to the existence of a substantial number of radiocarbon dates from early neolithic sites in Europe and the Near East. Another factor that made it feasible to attempt a rate determination was the availability of knowledge on the history of early cereal crops. As we saw in Chapter 2, wild cereals such as emmer wheat and barley that apparently had geographic distributions limited essentially to southwestern Asia were brought under domestication at sites in the Near East by 7000 B.C. and were subsequently introduced into Europe. It is also clear that by 6000 B.C. well-developed economic systems based on food production extended over most of the area from Iran to Greece. What was spreading, in fact, was a complex and highly successful economic system that usually involved domesticated plants and animals and permitted substantial increases in population density. In trying to account for the rise of population levels that occurred during neolithic times, it would appear that the important factor was an increase in fertility rates, linked with the emergence of a sedentary way of life, and perhaps the positive economic value of a larger number of children. As can be inferred from the study of contemporary populations of hunters and gatherers, low fertility rates and a relatively long spacing between successive births would seem to be characteristic of pre-agricultural populations.

When the rate of spread is measured, it turns out to be on average about one kilometer per year. In comparison with historical events that we are more familiar with, this rate seems to be very slow—it took some 2,500 years to go from Turkey to England—but in fact 25 kilometers per generation is not an insubstantial distance to be traveled, especially in the context of the means of transportation available in neolithic times. Some variation in the local rate of spread can be observed in the western Mediterranean and in the area occupied by the Bandkeramik culture. In the former case, there is indirect evidence for the use of boats, which may have contributed to the somewhat faster local rate. On the other hand, the rate seems to be slower in areas such as the Alps where ecological conditions are less well suited to early forms of agriculture. As more and more radiocarbon dates accumulate in Europe, we should be able to develop a more refined picture both of the overall pattern and of local variations in the rate of spread. There will also be the opportunity to learn more about the relationships between early farming populations and the latest hunters and gatherers in various regions of Europe as more archaeological surveys and excavations are done. It is of particular interest to establish the forms that such population interaction may have taken, ranging from acculturation to warfare, disease, and mutualism.

In attempting to interpret the spread of early farming, we have noted that two main modes of explanation need to be contrasted. The first of these, cultural diffusion, would see the innovation of farming as being passed from one group to the next without substantial movement of farming populations. According to the second explanation—demic diffusion, as we have called it—the spread of farming would derive from the local growth and expansion of neolithic populations. The kinds of movements required in the latter case may be simply those connected with the relocations of households or settlements over short distances to previously uncultivated areas. It is worth recalling that these two modes of explanation need not be regarded as mutually exclusive; their combination is indeed possible, and in some areas both are likely to have contributed to the process.

The observed pattern of spread suggests that it might be useful to consider a model originally proposed by R. A. Fisher. The wave of advance model, as it is called, predicts that a wave front will be set up in a population experiencing both growth and local migratory activity and that it will advance at a constant radial rate. The rate of advance can be predicted on the basis of the rates of population growth and migratory activity. The model thus offers a means of comparing the observed rate of advance with that expected if the diffusion is demic, provided that estimates can be obtained for the growth and migratory components.

Although the archaeological record gives a clear impression that changes in population levels are associated with the neolithic transition, there are at present only a few areas in Europe where direct, semi-quantitative estimates of neolithic density levels and rates of population growth can be made. It is only quite recently that archaeologists have become more interested in exploring population questions and developing methods for estimating demographic quantities. Our own fieldwork in the region of Calabria in southern Italy suggests that major strides can be made once the archaeologist begins to focus on such questions. We can expect that over the next generation substantial progress will be seen in this area. Much greater challenges are involved in obtaining estimates of migratory activity. At the present time, the main source of information along these lines is that provided by ethnography, where data of the kind required are still scarce. It is to be expected that a better understanding of migratory processes in general will emerge as small-scale human populations are studied more closely.

Using information that is available on growth and migratory activity, we have attempted to compare in Chapter 5 the observed rate of advance with that predicted by the wave of advance model for demic diffusion. The latter was based on a range of values for growth and migration rates obtained mostly from ethnographic sources. We found that the observed rate of advance is compatible with the expectations of the wave of advance model. We have used the word "compatible" intentionally here in order to emphasize that agreement with the demic model does not necessarily mean that the demic explanation is the only correct one.

The micro-simulation of the spread of the Bandkeramik culture presented in Chapter 7 is an example of how the abstract and general formulation of the wave of advance model can be translated into a more tangible representation. As we have seen in the simulation, only modest amounts of migratory activity are required to set up a wave front that advances at a more or less constant rate. This micro-simulation study should be viewed as a heuristic exercise, useful for giving some idea of how the spread occurs in the context of settlement patterns and also how further evaluation of the demic hypothesis might be developed along archaeological lines. Simulation work will also continue to be of importance for a better understanding of how the spread occurs in two dimensions. As mentioned in Chapter 5, Fisher's model was originally formulated in terms of a linear habitat. From an analytical point of view, its extension to two dimensions still represents a challenge. On the basis of simulation work, the major expectations for the wave of advance seem to hold at least approximately in two dimensions.

A full measure of how useful it can be to bring together archaeological and genetic thinking may be gauged by considering the genetic implications of the wave of advance model. To the population biologist it is evident that such a radiation is likely to produce a characteristic geographic pattern of genes. The examination of this pattern among European populations may provide a way of evaluating the relative importance of the two models, demic and cultural, for the spread of early farming. Conversely, archaeology provides the geneticist with specific scenarios of historical and prehistorical events that can help in the analysis and interpretation of observed genetic patterns. It is worth remembering that the present wealth of genetic data is a recent development and that the coverage of the world populations for various genes is still very uneven. The major step in interpreting the patterns has involved going beyond the consideration of a single gene to what is usually called the multivariate approach.

As we saw in Chapter 6, data on individual genes such as Rh− or ABO suggest that historical events might have played an important role in their geographic distribution. With individual genes, however, random noise and the operation of other mechanisms, such as natural selection, add considerable complexity to the interpretation. The need for a synthetic treatment becomes intuitively clear when one undertakes the examination of a series of maps for individual genes: on one hand, these tend to show highly idiosyncratic patterns; on the other hand, there are some common elements as well. Multivariate statistical techniques can help considerably by abstracting from the variation of each gene the part that

CONCLUSIONS

is due to causes of variation common to all genes. Among these, historical processes such as the neolithic transition are likely to play a central role. As seen in Chapter 6, plotting the contour maps of the leading principal component shows that the neolithic transition forms the backbone of the geographic distribution of genes in Europe. It should be added that the actual working out of a method for the construction and display of geographic maps of gene frequencies and of their principal components took some time and effort and involves technical aspects that have not been discussed in any detail in this book.

Some confidence in the results of the gene mapping work comes from the fact that similar results are obtained after subdivision of the genetic data into two major subsets: one for HLA and the other for the remaining genes. It is of interest that the geographic distribution of the second and third principal components of European genes suggests that other events of historical interest may be reflected as well. The macro-simulation briefly described in Chapter 7 indicates that the enmeshed and superimposed patterns of several major demic "events" can be sorted out by principal components analysis. The point should be made here that analyses of this kind are still at an early stage of development and that we can expect much more refined interpretations to be achieved with further investigation. Data on more populations and also on more genes are becoming available; more components beyond the third can be studied; more information of archaeological and other nature can assist in guiding the analysis. The other events shaping human populations may be somewhat more difficult to deal with than the neolithic transition, since they probably involved much smaller relative changes in levels of population density.

In addition to the broader genetic trends related to the neolithic transition, this "event" may also have affected individual genes in specific ways as a consequence of natural selection determined, for example, by dietary changes. In this regard, it is worth mentioning that recent work has shown a relation between intolerance to gluten, a protein contained in wheat and barley, and HLA-B8, a form of the HLA gene.[1] Another example would appear to be lactose absorption by adults, which shows a geographic distribution corresponding to the domestication of cattle.[2] When consumed in a fresh form, lactose, which is a sugar in milk, cannot be properly metabolized by ethnic groups other than those living in Europe and southwest Asia and also some parts of Africa.

Work remains to be done to assess what can be learned from

skeletal data about the neolithic transition. From the point of view of human biology, bones have the potential of providing us with information on populations at the time that the events of interest were taking place in the past. Unfortunately, they cannot now provide a direct means of obtaining genetic information. Perhaps the most promising lines of investigation, as seen in recent studies of skeletal remains, relate to such questions as nutrition, disease, and the timing of important events in an individual's life, such as the age at which a child is weaned.[3] Evidence along these lines may help to clarify various questions of demographic interest.

In this book we have been primarily concerned with the spread of early farming to Europe. It is obviously of equal interest to see if such a spread can be seen in other directions from the same center in southwestern Asia. The archaeological evidence available at present is still quite limited and allows us to form only a sketchy picture of developments in North Africa, the Arabian Peninsula, eastern Iran and Afghanistan, and the Caucasus. In the case of North Africa, the early economic complex of crops and animals found along the Mediterranean coast and in the lower Nile Valley seems to derive from the Near East. It is worth mentioning that in the Sahara, cattle pastoralism is documented at an early date (perhaps as early as 6000 B.C.) and may have developed independently. In the case of the Arabian Peninsula and eastern Iran there is local evidence for the arrival of agricultural economies at least by 4000 B.C. Unfortunately, the situation in the area between eastern Iran and the Indus Valley, where a major center of civilization emerged by 3000 B.C., is still not well known for the period between 6000 B.C. and 4000 B.C. North of the Caucasus the expansion of farming seems to date back to at least the fifth millennium B.C. Over the next twenty years archaeological investigations should make it possible to document more fully the pattern of spread in other directions as well as to obtain evidence for independent local experiments at food production in certain regions. For various reasons, the genetic information for these parts of the world is often still quite limited as well. Again, we expect that new information will become available in the future. The exception here will be those areas such as the Sahara that are no longer suitable for human occupation due to desertification and whose populations have in most cases become dispersed.

As we have seen in Chapter 2, there are several independent centers for the origins of food production in other parts of the world. It would seem well worth investigating the same set of re-

CONCLUSIONS

lationships between the origins and spread of agricultural systems and the patterns of gene frequency distributions in these other cases. Newly available information along archaeological lines may make this possible in the near future. It is worth adding that we may anticipate somewhat different scenarios than the one observed in Europe owing to different ecological conditions in other parts of the world. For the case of sub-Saharan Africa, one demic event of some magnitude was the spread of Bantu-speaking farmers to the southern part of Africa from a region presumed to have been located between Nigeria and Cameroon.[4] The genetic homogeneity of Bantu-speaking populations has been noted for some time.[5] In their expansion, Bantu farmers established essentially mutualistic relationships with populations of local hunter-gatherers that have in some cases survived up to the present. Perhaps the most promising parts of the world to consider in terms of the demographic consequences of food production will be those in East Asia and Central America.

What we have attempted is a reconstruction that brings together archaeological and genetic lines of information. In the case of both evolutionary and historial reconstructions, we are not dealing with phenomena that are occurring under controlled conditions and that can be repeated at will in the laboratory. The study of such phenomena represents a scientific challenge, and the degree of confidence that can be placed in conclusions drawn can seldom match that which is obtained in the experimental sciences. Confidence in reconstructions is built by the development of multiple lines of evidence that generate independent support for a particular interpretation. Ultimately, it is the growth of new evidence in individual fields and the creation of expectations for findings in other fields that generate a dense network for evaluating a reconstructive hypothesis. Its usefulness in the present case will have to be evaluated as new evidence becomes available and is tested against the argument that we have put forward. The remoteness of the events in which we have been interested, in combination with the time scale over which they unfold, makes them even more challenging to study. On the other hand, they may represent some of the most momentous transformations in the evolution of the human condition.

APPENDIX

The following lists give the geographic coordinates of the sites used to make the isochron maps in figures 4.5 and 4.6, and also the laboratory numbers of the samples providing C-14 dates. In almost all cases the coordinates are those cited in the description of a site in *Radiocarbon*. Unless otherwise indicated, the longitude is east of the prime meridian at Greenwich. An asterisk indicates that the coordinate is known only approximately. The letters preceding a sample number identify the laboratory (see the abbreviations in *Radiocarbon*). Determinations based on samples of bone are so indicated. Ertebølle sites are placed in Site List 4.2, except that those belonging to the most recent phase of the culture are included in Site List 4.1 (i.e., numbers 94, 95, and 96). In the Soviet Union, early sites belonging to the so-called forest neolithic cultures that have pottery but subsistence economies based on hunting, fishing, and gathering (i.e., without agriculture) are placed in Site List 4.2.

SITE LIST 4.1. The neolithic sites with C-14 dates used to make the isochron map in figure 4.5 and with geographic coordinates and laboratory numbers

Site	Date B.P.	Lat	Long	Sample Number
1. Aswad	9690*	33.36*	36.30*	GIF-2372, 2633
2. Jericho	9370*	31.85	35.45	BM-1321, 1322, 1323, 1324, 1326, 1327
3. Beidha	8630*	30.37	35.43	GRN-5063; P-1379; K-1082
4. Ali Kosh	8840	32.42	47.25	SHELL-1174 (see also UCLA-750D)
5. Ganj Dareh	9030*	34.33	47.50	P-1484, 1485, 1486
6. Çayönü	9360*	38.23	39.05	GRN-4458, 4459
7. Aşikli Hüyük	8790*	38.37	34.25	P-1238, 1239, 1240, 1241, 1245
8. Haçilar	8700	37.58	30.08	BM-127
9. Shomu Tepe	7510	41.12	45.45	LE-631
10. Togolok Depe	7320	38.12	57.95	BLN-719
11. Ghar-I-Mar	7125*	36.08	66.75	HR-428, 429
12. Tutkaul	7100	38.33	69.22	LE-690
13. Dhali-Agridhi	7990	35.03	33.43	P-2548; bone
14. Khirokitia	7595*	34.80	33.35	ST-414, 415, 416 (see also BM-854)
15. Kalavasos-Tenta	8350	34.75	33.30	P-2548
16. Knossos	7875*	35.52	25.33	BM-124, 278, 436
17. Franchthi	7795	37.42	23.08	P-1392 (see also P-1525)
18. Elateia	7480	38.62	22.73	GRN-2973
19. Sesklo	7755	39.38	22.82	P-1681 (see also P-1680, 1682)
20. Achilleion	7430*	39.30*	22.40*	LJ-3329, 4449; P-2118; UCLA-1896A
21. Argissa	7500	39.63	22.47	GRN-4145 (see also dates on bone: UCLA-1657A, 1657D)
22. Nea Nikomedia	8180	40.65	22.30	Q-655 (see also P-1212)
23. Sidari	7670	39.75*	19.75*	GX-771
24. Veluska	6950	40.90	20.35	TX-1785
25. Soroki	7515	48.18	28.32	BLN-588 (see also BLN-587)
26. Azmak	7030*	42.45	25.77	BLN-291, 292, 293, 294
27. Anza	7190*	41.70	20.00	LJ-2181, 2330, 3033, 3183, 3186
28. Grivac	7250	43.98	20.70	BLN-869
29. Banja	7050	44.18	20.57	BLN-730

30.	Divostin	6940*	44.03	20.83	BM-573; BLN-896 (see also BLN-823, 824, 826)
31.	Gornja Tuzla	6640	44.45	18.77	GRN-2059
32.	Desyk-Olajkut	6570*	46.22	20.25	BLN-581, 584
33.	Gyálarét	7090	46.22	20.08	BLN-75
34.	Katalzeg	6370	46.67	21.10	BLN-86
35.	Kotacpart	6450	46.42	20.32	BLN-115
36.	Scaramella	7000	41.50	15.68	R-350 (see also R-351)
37.	Redina	7110	41.25	16.00	LJ-4548 (see also LJ-4549)
38.	Piana di Curinga	6930	38.82	16.25	P-2946
39.	Grotta dell'Uzzo	6750	38.12*	12.50*	P-2733
40.	Penne	6580	42.47	13.93	PI-101
41.	Maddalena	6580	43.07	13.07	R-643
42.	Romagnano	6060	46.07	11.13	R-781 (see also R-780)
43.	Arene Candide	6680*	44.50	8.50	LJ-4143, 4144
44.	Le Capitaine	6050	43.75	6.17	GIF-1111
45.	St. Mitre	6550*	43.83	5.58	MC-263, 264 (see also MC-265)
46.	Châteauneuf	6430	43.38	5.17	LY-446 (KN-182 is unreliable; see Guilaine 1976)
47.	Montclus	6300	44.27	4.43	LY-303 (see also KN-181)
48.	Baratin	6600	44.25	4.88	GIF-1855
49.	Combe Obscure	6400	44.48	4.13	LY-423
50.	Le Suc	5980	44.85	3.92	LY-1057
51.	Camprafaud	6300	43.43	2.90	GIF-1491
52.	Gazel	6905	43.25	2.75	GRN-6702 (see also GRN 6705, 6706, 6707)
53.	Jean Cros	6600	43.08	2.47	GIF-3575 (see also GSY-218)
54.	Roucadour	5940	44.83	1.50	GSY-36A
55.	Coveta de l'Or	6445*	38.70	0.47 W	KN-51; H-1754
56.	Murciélagos	6275*	37.80*	4.85*W	GRN-6638, 6926 (see also CISC-53, 54, 55)

Note: These sites provided the estimates of time of arrival of early farming in various parts of Europe that were used in generating the isochron map shown in figure 4.5. The C-14 dates are conventional radiocarbon ages in years B.P. An asterisk indicates a case in which two or more C-14 dates are available for a site; these dates have been averaged.

SITE LIST 4.1, cont'd

	Site	Date B.P.	Lat	Long		Sample Number
57.	Sturovo	6215*	47.80	18.73		BLN-558, 559
58.	Zopy	6430	49.33	17.58		BLN-57
59.	Mohelnice	6345*	49.78	16.92		BLN-102, 102A
60.	Pulkau	6265	48.70	15.87		BLN-83
61.	Bylany	6185	49.92	15.33		BM-562 (see other dates by BLN, GRN, and LJ)
62.	Chabarovice	6400	50.67	13.95		BLN-438
63.	Olszanica	6430	50.10	18.83		GRN-5384
64.	Breść Kujawski	6180	52.60	18.58		GRN-9255 (see also LOD-172)
65.	Strezelce	6260	53.32	18.12		GRN-5087
66.	Zwenkau-Harth	6030*	51.23	12.35		GRN-1581; BLN-66; K-555
67.	Westeregln	6105*	51.97	11.40		GRN-223; BLN-43, 92
68.	Eitzum	6310	52.15	10.80		BLN-51
69.	Rosdorf	6350	51.50	9.90		HR-586
70.	Friedburg	6120	50.33	8.75		BLN-56
71.	Lautereck	6140	48.33	9.57		GRN-4750
72.	Evendorf	6050	45.42	6.57		LY-1181
73.	Reichtett	5940	48.63	7.75		LY-865
74.	Elsloo	6270	50.93	5.77		GRN-2164
75.	Geleen	6265*	50.97	5.83		GRN-995, 996
76.	Sittard	6095	51.00	5.87		GRN-320
77.	Swifterbant	5300*	52.57	5.58		GRN-6899, 7042, 7043, 7044
78.	La Hoguette	5560	49.10	0.37 W		LY-131
79.	Les Longrais	5290	48.65	0.22 W		LY-150
80.	Ile Carn 1	5230	48.58	4.68 W		GRN-1968
81.	Ile Carn 2	5390	48.57	4.68 W		GIF-1362
82.	Curnic	5340	48.63	4.47 W		GRN-1966
83.	Feldmeilen	5415	47.32	8.62		UCLA-1691F (see other dates by UCLA)
84.	Niederwil	4990	47.57	8.90		GRN-4203 (see also GRN-4202, 4204)
85.	Eaton Heath	5095	52.62	1.27		BM-770

86.	Fussell's Lodge	5180	51.08	1.73 W	BM-134
87.	Windmill Hill	5190	51.43	1.87 W	BM-180
88.	Hembury	5235*	50.82	3.27 W	BM-136, 138
89.	Lambourn	5365	51.50*	1.50*W	GX-1178
90.	Coygan Camp	5000	51.75	4.50 W	NPL-132
91.	Monamore	5110	55.52	5.13 W	Q-675
92.	Mad Man's Window	5095	54.97	5.92 W	UB-205
93.	Ballynagily	5625*	54.70	6.85 W	UB-197, 305, 307, 559
94.	Olby Lyng	5265*	55.50	12.22	K-1230, 1231
95.	Christianholm	5340*	55.75	12.57	K-729, 750
96.	Lietzow Buddelin 1	5190	54.48	13.38	BLN-560
97.	Ringkloster	5320	56.02	9.95	K-1654
98.	Konens Hoj	5260	55.98	10.67	K-923
99.	Lindebjerb	5010	55.70	11.18	K-1659
100.	Solager	4650	55.93	11.90	K-1724
101.	Hagestad	4700	55.40	14.18	LU-1349
102.	Hjulberga	4830	59.35	15.12	LU-1319
103.	Sarnowo	5570	52.48	18.75	GRN-5035
104.	Sventoji	4160*	55.95	21.25	LE-833, 904
105.	Sārnate	4500*	57.13	21.43	LE-814; TA-24 (see also BLN-769)
106.	Piestiné	4595*	56.92	27.00	LE-748, 750

SITE LIST 4.2. The mesolithic sites with C-14 dates used to make the isochron map in figure 4.6 and with geographic coordinates and laboratory numbers

	Site	Date B.P.	Lat	Long		Sample Number
1.	Abu Hureyra	10,790	35.87	38.40		BM-1121
2.	Mureybat	10,040*	36.20	38.12		P-1215, 1216, 1217, 1220, 1222 (see also LV-605, 606, 607)
3.	Nahal Oren	10,045	32.67	35.00		BM-764; bone
4.	Jericho	11,090	31.88	35.45		BM-1327
5.	Rosh Horesha	10,650*	30.52	34.52		SMU-9, 10
6.	Abu Salem	9970	30.52	34.55		I-5498
7.	Ganj Dareh	10,400	34.33	47.25		GAK-807
8.	Machai	7550	38.25	67.25		LE-982
9.	Franchthi	7960*	37.43	23.13		P-1526, 1527
10.	Sidari	7820	39.75	19.75		GX-?
11.	Vlasac	7615*	44.53	22.05		LJ-2407; Z-267
12.	Lepenski Vir	6860*	44.52	22.03		BM-379; P-1598
13.	Odmut	7350	45.20	18.83		Z-413
14.	Benussi	7050	45.78	13.70		R-1043
15.	Covoloni del Broion	6930	45.47	11.38		R-892
16.	Romagnano	6480	46.07	11.13		R-1136
17.	Pradestel	6870	46.12	11.08		R-1148
18.	Grotta dell'Uzzo	7910	38.12	12.50		P-2734
19.	Châteauneuf	7270	43.38	5.17		LY-448 (see also LY-438)
20.	Montclus	6230	44.27	4.43		LY-494 (see also LY-495, 496)
21.	Longetraye	6210	44.87	3.92		LY-616
22.	Solignac	7100	44.93	3.90		LY-539
23.	Montbani	7280	49.50	1.25		GIF-355
24.	La Torche	5970	47.83	4.28	W	GRN-2001
25.	Cabeço da Armoreira	6050	38.83	8.92	W	SA-194
26.	Birsmatten	5350	47.43	7.55		B-234; bone (see also B-235)
27.	Lautereck	6440	48.33	9.57		GRN-4667
28.	Maarheeze	6230	51.28	5.63		GRN-2446
29.	Tilburg-Labe	6500	51.58	5.05		GRN-1597
30.	High Rocks	5695*	51.12	0.23		BM-40, 90
31.	Wakeford Copse	5680	50.88	0.97	W	HAR-233
32.	Wawcott	5260	51.40	1.43	W	BM-449 (see also BM-767, 826)

33.	Dumford Bridge	5380	53.50	1.80 W	Q-799
34.	Rocher Moss	5830	53.42*	1.83*W	Q-1190
35.	March Hill	5850	53.62	1.97 W	Q-788 (see also Q-1188)
36.	Cnoc Sligeach	5425	56.02	6.22 W	BM-670 (see also GX-1903, 1904)
37.	Caisteal-nan-Gillean	5450	56.02	6.32 W	BIRM-347 (see also BIRM-347, 348)
38.	Cnoc Coig	5465*	56.00*	6.30*W	Q-1351, 1352 (see also Q-1353, 1354)
39.	Sutton	5250	53.38	6.08 W	I-5067
40.	Dalkey Island	5300	53.28	6.08 W	D-38
41.	Ringneill Quay	5380	54.53	5.50 W	Q-770
42.	Lough Derravaragh	5360	53.67	7.37 W	I-4234
43.	Lietzow Buddelin 2	5815	54.48	13.53	BLN-561
44.	Ralswick-Augustenhof	5455	54.48	13.48	BLN-562
45.	Kiel-Ellerbek	5170	54.37	10.18	KI-152
46.	Ringkloster	5610	56.02	9.95	K-1652
47.	Flynderhage	5230	56.02	10.23	K-1450
48.	Norslund	5705*	56.02	10.23	K-990, 991
49.	Solager	5520	55.93	11.90	K-1723
50.	Hojelse	6080	55.47	12.20	K-1098
51.	Salpetermose	5690*	55.92	12.30	K-1232, 1233, 1234, 1235
52.	Arlov	6225*	55.62	13.07	LU-756, 757
53.	Skateholm	5790	55.38	13.48	LU-1848
54.	Torsröd	5350	59.00	10.33	T-1425
55.	Hein	5085*	60.37	7.73	T-1001, 1005
56.	Gyrinos	5700	60.75	8.20	K-711
57.	Zatsen'ye	5450	54.00*	27.00*	LE-960
58.	Osa	5780	57.00*	27.00*	LE-962 (see also LE-850, 961)
59.	Narva	5300	59.33	28.02	TA-7 (see also TA-33)
60.	Zarech'ye	5670	56.00*	37.00*	LE-969
61.	Ivanovskoye	5730	57.00*	38.50*	GIN-241
62.	Berendeyevo	5730	57.50*	39.00*	GIN-112

Note: These sites provided the estimates of the time of latest mesolithic occupation in various parts of Europe that were used in generating the isochron map shown in figure 4.6. The C-14 dates are conventional radiocarbon ages in years B.P. An asterisk indicates a case in which two or more C-14 dates are available for a site; these dates have been averaged.

NOTES

Chapter 1. Introduction

1. Coale 1974; Hassan 1981.
2. Zohary 1969. So far, the Franchthi Cave in Greece is the only known site in Europe where cereals (barley and oats) have been recovered from mesolithic contexts (Hansen and Renfrew 1978).
3. Renfrew (1973) has been the leading advocate of the recent emphasis on endogenous cultural developments within Europe. A good case in support of this position would be the autonomous development of early metallurgy in the Balkans.

Chapter 2. The Origins of Agriculture

1. In Asia, claims have been made for the early appearance of domesticated forms of legumes, but not cereals, at the Spirit Cave in Thailand (Gorman 1971). In the Nile Valley, cereal remains have been recovered from contexts at Wadi Kubbaniya that appear to date back to well before 12,000 B.C. (Wendorf and Schild 1980b); recent C-14 determinations done directly on the seeds indicate that they are much younger in date and may be only of Classical age.
2. Chang 1981.
3. Flannery 1973.
4. McIntosh and McIntosh 1982.
5. Wendorf and Schild 1980a.
6. See, for example, the contributions by Harris and Wright in Reed 1977.
7. Zohary 1969.
8. Geist 1971.
9. Bökönyi 1976.
10. Harlan 1975.
11. Van Zeist and Bottema 1981.
12. Bökönyi 1976; Clutton-Brock 1981.
13. Renfrew et al. 1968.
14. As Moore (1982) has recently argued, previous models, such as those of Flannery and Binford, do not seem to stand up well in the light of the evidence recovered from excavations at Tell Aswad, Abu Hureyra, and Mureybat in the 1970s. A new generation of models focusing on developments that occurred in the ninth millennium B.C. (i.e., at the end of the Pleistocene) needs to be explored.

Chapter 3. The Neolithic Transition in Europe

1. Daniel 1967.
2. Examples include Ertebølle pottery in Denmark and the comb decorated pottery of the so-called forest neolithic cultures of the northwestern part of the Soviet Union (Dolukhanov 1979).
3. See, for example, Startin 1978.
4. Fairbridge and Hillaire-Marcel 1977.
5. Bay-Peterson 1978.
6. In the case of Lepenski Vir, there is some discussion in the literature about the attribution of various parts of the site, including the structural remains, to the mesolithic period. See, for example, Milisauskas 1978: 96.
7. There is now some evidence to support the argument developed by Clarke. Various seed remains, including those of legumes (lentils and peas), have been recovered from mesolithic layers at the sites of Balma Abeurador and Fontbrégoua in southern France.
8. Kozlowski 1973; Mellars 1978; Clark 1980.
9. For those less familiar with neolithic studies in Europe, the following books provide useful regional syntheses: the Balkans (Tringham 1971), the western Mediterranean (Phillips 1975), central Europe (Milisauskas 1978), and the Soviet Union (Dolukhanov 1979).
10. Soudsky and Pavlu 1972; Milisauskas 1978: 76-80.
11. Kuper et al. 1974; Kuper and Lüning 1975.
12. Our own work in Calabria provides a good example of the situation. Until 1974, neolithic occupation in Calabria (the region comprising the toe of southern Italy) was known essentially from three caves—Grotta della Madonna, Romito, and Grotta di Sant'Angelo—located in the rugged northern part of the region. Surveys and excavations since 1974 have revealed dense patterns of open-air settlement of impressed ware neolithic date in areas of Calabria well suited to early farming (Ammerman and Shaffer 1981).
13. The early C-14 dates from the sites of Cap Ragnon and L'Ile Riou in southern France are based upon the dating of shell samples. There is some question about the reliability of dates obtained from this material, which appear to be too old (Guilaine 1979a: 209-215). On Corsica, there are several early dates from the site of Curacchiaghiu, but excavations have apparently produced only four pottery sherds. A single early date (GIF-1961) was also obtained for the site of Basi from levels with cardial impressed pottery (see Phillips 1975; Weiss and Lanfranchi in Guilaine 1976). Most of the C-14 dates from southern France based upon charcoal samples fall in the fifth millennium B.C. (uncalibrated; Delibrias et al. 1982). See also the discussion of J. Evin on the revision of the absolute chronology for the beginning of the neolithic in Provence and Languedoc, to appear in the forthcoming volume, *Premières communautés paysannes en Méditerranée occidentale*. Also, several cave sites in Spain—Cueva del Parco, Cova Fosca,

NOTES TO CHAPTER 4

Cueva Matutano, and Cova con Ballester—have produced rather early C-14 dates in comparison with the pottery styles present, and those dates have been the subject of considerable debate.

14. Geddes 1981.
15. Cherry 1981; Ammerman 1979.
16. Randsborg 1975.
17. The one exception here would seem to be the Franchthi Cave in Greece, where barley is present in mesolithic contexts; emmer wheat seems to make its appearance only in early neolithic levels (Hansen and Renfrew 1978). No wild or domesticated forms of wheat or barley appear to be represented among the abundant plant remains recovered from the mesolithic levels at Balma Abeurador and Fontbrégoua in southern France. Even those who try to stress continuity in the transition from the mesolithic to the neolithic accept the introduction of domesticated cereals into Europe (e.g., Clark 1980: 64-67). A good case can also be made for the introduction of sheep and goats into such areas as the Hungarian plain (Bökönyi 1974), southern France (Geddes 1981), and southern Italy (the remains of sheep and goat are not seen in the mesolithic sequences at the Grotta della Madonna and Grotta dell'Uzzo, for example, and yet are well represented in neolithic levels at these two sites).
18. It is worth noting several exceptions to this pattern: early aceramic neolithic occupation in Greece (as seen, for example, in the lowest levels at Knossos); early cardial pottery possibly preceding the initial practice of farming in southern France; and Ertebølle pottery in the Baltic area. The use of pottery also precedes the appearance of early forms of agriculture in northwestern parts of the Soviet Union.
19. One of the limitations of site catchment studies (see various contributions in Higgs 1975), where the soils near a site are classified in terms such as "arable," "grazing," and "non-arable," is that the classification system derives from a market-oriented view of production rather than one emphasizing subsistence. As seen in our own work near Acconia in Calabria, there may well have been a preference for lighter soils that have lower yields but are much easier to work.
20. See, for example, Kruk 1980 and Halstead 1981.
21. Dennell 1979. A comparatively new line of investigation that appears to hold some promise is the chemical analysis of human skeletal remains. Tauber (1981), for example, documents a shift from reliance on the exploitation of marine resources at mesolithic sites along the coast of Denmark to dependence on terrestrial foods at neolithic sites in the same region.
22. Bender 1978.

Chapter 4. Measuring the Rate of Spread

1. It is desirable not to base a rate measurement upon time differences with respect to a starting date for the diffusion process, as Edmonson

(1961) does, since it is likely to be difficult to pinpoint such a "starting time." However, there are many points in space at some distance from the origin of the process where one can potentially obtain estimates of "time of first arrival" (Ammerman and Cavalli-Sforza 1971: 677).

2. There is an extensive literature on the relationship between C-14 dates and calendric dates, as well as on the construction of calibration curves (see the references in Olsson 1970; Clark 1979; Klein et al. 1982). A graphic representation of the basic trends in this relationship over the last 8,000 years is given in figure 4.4. This representation summarizes the recently published "Tables based on the consensus data of the Workshop on Calibrating the Radiocarbon Time Scale" (Klein et al. 1982), which derives from 1,154 samples of dendrochronologically dated wood analyzed by the radiocarbon laboratories at the Universities of Arizona, Groningen, California at La Jolla, Pennsylvania, and Yale. It is conventional for radiocarbon dates to be reported in years B.P., or before present (0 B.P. being taken as 1950 A.D.), using the shorter Libby (5,568-year) half-life for C-14. In the case of radioactive decay, there is a Poisson error associated with an activity measurement, and thus a standard error is reported for each radiocarbon date. Uncalibrated C-14 dates are used in the regression analysis (figure 4.2) and the two isochron maps (figures 4.5 and 4.6), since the data sets include dates before 7200 B.P. and calibration tables at present do not extend back before this date.

3. From a statistical point of view, there is a choice between at least three straight lines that could be fitted to the data: the two regressions (distance on time and time on distance) and the principal axis that passes approximately midway between the two regression lines (see Ammerman and Cavalli-Sforza 1971: fig. 2). The first regression would be preferable if most of the error in measurements of the data was due to dating (here the magnitude of the error is approximately known from the standard errors associated with C-14 dates), and the second if it was due to estimated distances. The distances are in appearance exact, but because of geographic uncertainties, great circle routes in fact involve an approximation whose associated error is difficult to estimate. When there is doubt over the relative importance of the two sources of error, the slope of the principal axis is preferred as a means of measuring the average rate of spread (Ammerman and Cavalli-Sforza 1971: 681). See Sokal and Rohlf 1981 on how it is computed.

4. Earlier versions of the isochron map of the spread of early farming in Europe were published by Cavalli-Sforza (1974) and Menozzi et al. (1978). The 1974 map was based on polynomial surfaces fitted by least squares. The algorithm used for generating the isochron maps given by Menozzi et al. is described by Piazza et al. 1981a. A similar

NOTES TO CHAPTER 5

approach with more data was followed for figure 4.5. It is worth noting the basic correspondence between figure 4.5 and the more intuitively drawn map of the spread recently published in *The Times Atlas of World History* (Barraclough 1979: 42-43).

5. See Newell 1970 for an attempt to explore this question in terms of lithic remains in southeastern Holland.

CHAPTER 5. THE WAVE OF ADVANCE MODEL

1. The ethnographic data show considerable variation in densities of hunters and gatherers, depending upon the environment in which they live. Australian aborigines, living mostly in arid, semitropical areas, have the lowest densities in the world (0.01–0.05 persons per square kilometer). Also low are those of north Canadian Indians living inland (Chipewyan, Montagnais, and Algonkian: 0.016–0.12). By contrast, shoreline areas show relatively high densities. In fact, probably the world's highest population densities of food collectors are on the Alaskan and Canadian coasts (Haida: 9.5 persons per square kilometer). This is because the catch of large fish in estuaries and certain rivers can be especially abundant. The median density for the shoreline and river fishers of the Pacific Coast of North America is approximately 1 person per square kilometer. These high densities are in reality limited to relatively small habitats in comparison with the total area of the continent. Also high were the densities of precolonization hunter-gatherers in California, who had developed elaborate techniques of food preparation not normally seen among hunters and gatherers (0.4–3.9). African hunters have to date densities ranging from 0.06 (bushmen) to 0.5–1.0 (Ituri pygmies). Altogether, a median value for hunter-gatherers is around 0.1–0.2 persons per square kilometer. References to this information, with the exception of the pygmies, can be found in Hassan (1981). These values are about the same as the estimates from archaeological data on hunter-gatherers of final paleolithic date in North Africa (Hassan 1981).
2. Acsadi and Nemeskeri 1970; Weiss 1973.
3. An estimate for modern bushmen suggests a population growth rate slightly above zero. This figure is based on a small and possibly biased sample (Howell 1979).
4. Lee 1972.
5. McKeown 1976.
6. Hassan 1981: 130-133.
7. In the model as adapted by Skellam to the spread of a population, the population grows according to a logistic curve that is described in section 5.4, whereby the number of individuals—or the local population density p—increases over time with a rate equal to:

$$\frac{dp}{dt} = ap(1 - p/M) \tag{1}$$

where a is the initial growth rate of the population and M is the maximum number (or density) that can be reached at saturation. At the same time, migratory activity is described by the classical diffusion equation:

$$\frac{dp}{dt} = m\frac{d^2p}{dx^2} \tag{2}$$

where x is a (unidimensional) spatial coordinate and m is a coefficient measuring the rate of migration. This treatment of diffusion is the analogue in continuous space of the random walk process described in the text. Diffusion according to (2) leads to the migration distribution of Gaussian or normal type. The two components—growth (in time) and migration (in space and time)—can be summed together:

$$\frac{dp}{dt} = ap(1 - p/M) + m(d^2p/dx^2). \tag{3}$$

It was shown by Fisher that p forms a wave that advances in space with constant shape and velocity (as shown in figure 5.2) and with a radial velocity (rate of advance) not less than $2\sqrt{ma}$.

8. From Ablowitz and Zeppetella (1979) one can obtain the following solution for Fisher's equation:

$$p = \{1 + \exp[(x - \rho t)/\sqrt{6}]\}^{-2}; \quad \rho = 5/\sqrt{6}$$

where x is distance from the origin and t is time.
We thank Dr. K. H. Chen for providing this reference.

9. Skellam 1973.

10. The rate of growth per unit time in the logistic model is given by equation (1) in note 7 above. Starting at $t = t_o$ with $p = p_o$, we have at time t:

$$p_t = [1 + (1 - p_o)e^{-at}/p_o]^{-1}$$

11. Information on the number of houses and the period of occupation on the Aldenhoven Platte comes from Kuper et al. 1974. In fact, the authors consider that 35% of the landscape was lost to examination and estimate that a total of 246 houses originally stood at one time or another in the investigated area. In a publication (Ammerman and Cavalli-Sforza 1979: 283-284) prepared prior to the time that the report by Kuper et al. 1974 was available to us, we attempted some trial estimates of population density for the area. Our approach in terms of assumptions about the life of a house (i.e., 25 years) was quite similar to that adopted by Kuper et al. 1974. To be on the conservative side, we took the total period of Bandkeramik occupa-

NOTES TO CHAPTER 5

tion to be 700 years rather than 500 years, since much higher local densities—as high as 5 persons per square kilometer at times—would obtain under the latter estimate for the length of the period.

12. In fig. 5.6, there are 11 houses at saturation, and so the first house is at $1/11 = 9\%$ of the saturation level. The last-but-one house during the growth process is at $10/11 = 91\%$ of the saturation level, which occurs some 250 years after the beginning of occupation in the area. We therefore put in the formula for logistic growth (see note 10 above) the following values: $p_t = 0.91$, $p_o = 0.09$, and $t = 250$. From $0.91 = 1/[1 + (1 - 0.09)e^{-250a}/(0.09)]$ we find that a, the initial growth rate, is 0.019, or 1.9% per year. This is an active growth rate even by the standards of modern populations. Given the small number of points on the curve, this heuristic analysis is obviously quite rough.
13. Skellam 1973.
14. Hewlett et al. 1982.
15. The Majangir, a group of shifting agriculturalists in the west of Ethiopia, have been studied by Stauder (1971), who has analyzed their mobility in terms of the distance between birthplace and place of residence for two communities that were censused. We limited our calculations here to age group 10-29 for the Gilishi community and to age group 10-19 for the Shiri community, which Stauder thinks correspond to the mobility in one generation. Older age groups show much greater apparent mobility, probably because they were chased from their earlier territory by raids of foreign tribes. Using central values for age and also for the distance classes, we find:

	Gilishi		Shiri	
	15 year old age group	25 year old age group	15 year old age group	Average
Mean distance (km)	21.2	26.1	42.7	30.0
Mean square distance (km^2)	1115.7	1325.6	2153.0	1531.4

Note that mean square distance is greater than the square of the mean distance (they could be equal only if no variation occurred in individual mobility). Stauder considers that Gilishi mobility is lower because movement by group members was discouraged. In general, these values are all somewhat higher than those obtained from mating distances.

16. To reduce this rate to an estimate per year, we note that the length of a generation in a rural European community examined over the last 300 years has been approximately 31 years for males and 25 years for females, averaging 28 years for the two sexes. Perhaps this quantity was somewhat smaller in neolithic times, with death occurring

earlier on average. Twenty-five years is probably a better estimate of the average generation length during early neolithic times. The quantity m is dimensionally similar to a diffusion coefficient in physics: the distance traveled by diffusion is proportional to the square root of time, so that the square distance is proportional to time and the mean square distance traveled per year is then in the range of $300/25 = 12$ km^2/year and $2{,}000/25 = 80$ km^2/year.

17. The curves in figure 5.9 are calculated on the basis of the equation: $a = 6r^2/M$, derived from $r = 2.04\sqrt{ma}$ and the relationship $M = 25m$. M is the mean square distance (km^2) per generation, m that per year; 25 is the assumed generation time in years; r is the rate of advance in kilometers per year; and a is the growth rate per year.

Chapter 6. The Analysis of Genes

1. Mourant 1954.
2. Levine and Stetson 1939.
3. We put "gene" in quotation marks here because there might be more than one gene involved. Many different forms of the Rh gene in fact exist: they make only slightly different compounds to be found on the surface of the red cells and are distinguished by suffixes such as R_1, R_2, R_o, and so forth. Whenever a person inherits a capital R gene (from even only one of the two parents), that person is Rh positive.
4. The proportion of Rh− genes and that of Rh− individuals are not the same, since it takes two Rh− genes (one of paternal and one of maternal origin) to make the negative individual. There are two types of Rh+ individuals: those that inherit an Rh+ gene from both father and mother (homozygotes) and those who inherit an Rh+ gene from one and an Rh− gene from the other (heterozygotes). Their proportions in the population are given by the Hardy-Weinberg laws: if p and q (where $p + q = 1$) are the respective proportions of Rh+ and Rh− genes, then those of Rh+ homozygotes are p^2, Rh+ heterozygotes are $2pq$, and Rh− individuals are q^2. Thus, if Basques have 45:55 proportions (or gene frequencies) of Rh+ and Rh− genes, the Rh+ homozygotes will be 20.25% heterozygotes 49.50% and Rh− 30.25%.
5. Kimura 1968.
6. Sgaramella-Zonta and Cavalli-Sforza 1973.
7. Menozzi et al. 1978.
8. Putting the values of x_A and x_B as the abscissa and ordinate of a Cartesian diagram, the four populations are represented by four points (white circles). The dark circles are the projection of the four points on the oblique straight line that is drawn as closely as possible to the set of four white circles. If we replace a white circle (the observed value) with the corresponding dark circle (its projection on the straight line), we lose some information on the populations. The

NOTES TO CHAPTER 7   157

information lost is represented by segments between each white circle and its corresponding dark circle. But the computation of the coefficients (i.e., 1.5 and -4) is made so that this loss is as small as possible. Moreover, the actual loss can be measured as a percentage of the total variation, and thus we can evaluate the proportion of information that is not accounted for by this procedure.

9. Piazza et al. 1981a.
10. Menozzi et al. 1978; Piazza et al. 1980; Piazza et al. 1981b.
11. Sokal and Menozzi 1982.
12. Zvelebil 1980.
13. Gimbutas 1970 and 1973. However, Goodenough 1970 places the center of origin of Indo-Europeans somewhat further to the west. The area appearing as a center of origin of gene expansion in figure 6.12 is somewhat intermediate between the regions favored by these two investigators.
14. Kidd 1980.

CHAPTER 7. SIMULATION STUDIES

1. Sgaramella-Zonta and Cavalli-Sforza 1973; Rendine et al. 1984.
2. A preliminary report on this simulation work is given in Ammerman and Cavalli-Sforza (1973b; for a brief description of some aspects of this work, see also Ammerman and Cavalli-Sforza 1979). The model of settlement occupation that we adopted in our initial simulation studies followed that of Soudsky and Pavlu (1972). They would see the occupation, abandonment, and subsequent reoccupation of settlements as linked with a rotating swidden system of farming.
3. The results of extensive work on the Aldenhoven Platte (Eckert et al. 1971; Eckert et al. 1972; Farruggia et al. 1973; Kuper et al. 1974; Kuper and Lüning 1975) have led to the proposal of an alternative view of the occupation of Bandkeramik settlements: that they were more permanent individual farmsteads that were periodically rebuilt at a site. In our more recent (unpublished) simulation studies, we have begun to explore models along these lines. It is worth drawing attention to major work in the field on Bandkeramik settlement patterns, as well as their interpretation, in Holland (e.g., Modderman 1970) and Poland (e.g., Kruk 1980). Brief comment should also be made here on a recent simulation study of Bandkeramik settlement patterns by Hammond (1981), which unfortunately is seriously flawed by the choice of an exponential model of growth and the use of a growth rate of 3% (see the discussions in section 5.3).
4. For an introduction to the Lotka-Volterra equations, see Roughgarden 1979.
5. Sgaramella-Zonta and Cavalli-Sforza 1973; Rendine et al. 1984.
6. Ethnographic values for hunter-gatherers are probably somewhat higher than those used here.

7. It might be argued that drift cannot be the only source of variation. However, if selection also contributed to the generation of genetic patterns, it would often be compatible with the maintenance of extensive similarity between neighboring populations. Such would be the case if a selective condition favored one form of a gene in one region and another in a distant one. By assuming, as we did, a completely random pattern of variation between the populations in the cells, we have probably made our final task more difficult. In future simulation studies it will be worth experimenting with patterns of selection that vary geographically.
8. The value of m was selected on the basis of ethnographic evidence. The migration rate used for farming populations ($m = 0.04$), when translated into a value in terms of square kilometers, is to be multiplied by 160^2 (160 km being the length of the side of each cell) and is thus equivalent to about 1,000 square kilometers. This figure is well within the range of ethnographic values given in Chapter 5.
9. This is the older of the two subsequent demic events indicated respectively by the second and third principal components of the genetic analysis of populations in Europe (figures 6.11 and 6.12). Its origin would be located slightly above the Black Sea.
10. As is commonly done, random genetic drift is simulated by straight binomial sampling (with repetition) of the genes present. The effective population sizes used for drift were one-third of the census sizes given in the text (see Cavalli-Sforza and Bodmer 1971).
11. Piazza et al. 1981b; Menozzi et al. 1984.

Chapter 8. Conclusions

1. Simoons (1981: fig. 4) observes that there is a negative correlation between the frequency of antigen HLA-B8 and the length of time that wheat farming has been practiced in various parts of Europe. He puts forward the hypothesis that low HLA-B8 frequencies may be attributed to selection against this gene because of the celiac syndrome (i.e., gluten intolerance). It is worth noting that celiac syndrome is a relatively rare disease and that it may be difficult to assess its impact in lowering B-8 gene frequencies.
2. It is known that lactose absorption in adults is a practice observed only in areas where there has apparently been a long tradition of the consumption (by adults) of fresh milk (i.e., lactose has not been converted to other nutrients). A summary of the work on this question is provided by Simoons (1978). The distribution of lactose absorption by adults is limited largely to those areas of the world that witnessed the spread of early farming from southwest Asia. Certain populations in Africa would be the exception here. At the same time, it should be noted that the genetic basis of lactose absorption is far from being clearly understood. Moreover, it is possible that the north-to-south

NOTES TO CHAPTER 8

gradient of lactose malabsorption observed in Europe may be related to another phenomenon: a vitamin-D sparing action of lactose utilization (Durham 1984).
3. See, for example, Sillen and Kavanagh 1982.
4. Bouquiaux 1981.
5. Hiernaux 1974.

BIBLIOGRAPHY

Ablowitz, M., and A. Zeppetella. 1979. Explicit solutions of Fisher's equation for a special wave speed. *Bulletin of Mathematical Biology* **41**:835-840.
Acsadi, G., and J. Nemeskeri. 1970. *History of human life span mortality.* Budapest: Akademiai Kiado.
Ammerman, A. J. 1975. Late Pleistocene population dynamics. An alternative view. *Human Ecology* **3**:219-233.
Ammerman, A. J. 1979. A study of obsidian exchange networks in Calabria. *World Archaeology* **11**:95-110.
Ammerman, A. J., and L. L. Cavalli-Sforza. 1971. Measuring the rate of spread of early farming in Europe. *Man* **6**:674-688.
Ammerman, A.J., and L. L. Cavalli-Sforza. 1973a. A population model for the diffusion of early farming in Europe. In *The explanation of culture change*, ed. C. Renfrew. London: Duckworth.
Ammerman, A. J., and L. L. Cavalli-Sforza. 1973b. A simulation study of Bandkeramik settlement patterns. Paper given at the Symposium on the Applications of Computer Simulation to Archaeology. Society for American Archaeology, San Francisco, May 1973.
Ammerman, A. J., and L. L. Cavalli-Sforza. 1979. The wave of advance model for the spread of early farming. In *Transformations: Mathematical approaches to culture change*, ed. C. Renfrew and K. L. Cooke. New York: Academic Press.
Ammerman, A. J., and G. D. Shaffer. 1981. Neolithic settlement patterns in Calabria. *Current Anthropology* **22**:430-432.
Ammerman, A. J., D. K. Wagener, and L. L. Cavalli-Sforza. 1976. Toward the estimation of population growth in Old World prehistory. In *Demographic anthropology: Quantitative approaches*, ed. E. Zubrow. Albuquerque: University of New Mexico Press.
Angel, J. L. 1972. Ecology and population in the Eastern Mediterranean. *World Archaeology* **4**:88-105.
Bahuchet, S. (ed.). 1979. *Pygmees de Centrafrique.* Paris: Selaf.
Bakels, C. C. 1978. *Four Linearbandkeramik settlements and their environments.* Analecta Praehistorica Leidensia 11. The Hague: Leiden University Press.
Bar-Yosef, O. 1980. Prehistory of the Levant. *Annual Review of Anthropology* **9**:101-133.
Bay-Peterson, J. 1978. Animal exploitation in Mesolithic Denmark. In *The early postglacial settlement of northern Europe*, ed. P. Mellars. London: Duckworth.

Behrens, H. 1973. *Die Jungsteinzeit im Mittlelbe-Salle Gebiet.* Berlin: VEB Deutscher Verlag der Wissenchaften.
Bender, B. 1978. Gatherer-hunter to farmer: A social perspective. *World Archaeology* **10**:204-222.
Biagi, P. 1980. Some aspects of the prehistory of Northern Italy from the final palaeolithic to the middle neolithic: A reconsideration of the evidence available to date. *Proceedings of the Prehistoric Society* **46**:9-18.
Binford, L. R. 1968. Post-Pleistocene adaptations. In *New perspectives in archaeology*, ed. S. R. Binford and L. R. Binford. Chicago: Aldine.
Bodmer, W. F., and L. L. Cavalli-Sforza. 1971. Variation in fitness and molecular evolution. Vol. 5: Darwinian, New Darwinian and Non Darwinian Evolution. *Proceedings of VI Berkeley Symposium on Mathematical Statistics and Probability.* Berkeley: University of California Press.
Bodmer, W. F., and L. L. Cavalli-Sforza. 1976. *Genetics, evolution and man.* San Francisco: Freeman & Co.
Bökönyi, S. 1974. *History of domestic mammals in central and eastern Europe.* Budapest: Akademiai Kiado.
Bökönyi, S. 1976. Development of early stock rearing in the Near East. *Nature* **264**:19-23.
Bordaz, J. 1970. *Tools of the old and new stone age.* Garden City, N.Y.: Natural History Press.
Boserup, E. 1965. *The conditions of agricultural growth.* Chicago: Aldine.
Bouquiaux, L. (ed.). 1981. *L'Expansion bantoue.* Vol. 3. Paris: Selaf.
Braidwood, R. J., and B. Howe. 1960. *Prehistoric investigations in Iraqi Kurdistan.* Studies in Ancient Oriental Civilization, no. 31. Chicago: University of Chicago Press.
Braidwood, R. J., and C. A. Reed. 1957. The achievement and early consequences of food production: A consideration of the archaeological and natural-historical evidence. *Cold Spring Harbor Symposia on Quantitative Biology* **22**:10-31.
Candolle, A. de. 1884. *Origin of cultivated plants.* London: Kegan Paul.
Cauvin, J. 1977. Les fouilles de Mureybet (1971-74) et leur signification pour les origines de la sédentarisation au Proche-Orient. *Annual of the American Schools of Oriental Research* **44**:19-48.
Cavalli-Sforza, L. L. 1963. The distribution of migration distances: models and applications to genetics. In *Human displacement.* Monaco: Entretiens de Monaco en Sciences Humaines.
Cavalli-Sforza, L. L. 1974. The genetics of human populations. *Scientific American* **231** (3):80-89.
Cavalli-Sforza, L. L., and W. F. Bodmer. 1971. *The genetics of human populations.* San Francisco: Freeman.
Chang, K. C. 1981. In search of China's beginnings: New light on an old civilization. *American Scientist* **69**:148-160.
Cherry, J. F. 1981. Pattern and process in the earliest colonization of the Mediterranean islands. *Proceedings of the Prehistoric Society* **47**:41-68.

BIBLIOGRAPHY

Childe, V. G. 1929a. *The most ancient east.* New York: Knopf.
Childe, V. G. 1929b. *The Danube in prehistory.* Oxford: Clarendon Press.
Clark, J. D. 1976. Prehistoric populations and pressures favoring plant domestication in Africa. In *Origins of African plant domestication*, ed. J. Harlan, J. De Wet, and A. Stemler. The Hague: Mouton.
Clark, J.G.D. 1965. Radiocarbon dating and the expansion of farming culture from the Near East over Europe. *Proceedings of the Prehistoric Society* **31**:57-73.
Clark, J.G.D. 1980. *Mesolithic prelude.* Edinburgh: Edinburgh University Press.
Clark, R. M. 1979. Calibration, cross-validation and carbon-14. *Journal of the Royal Statistical Society A* **142** (1):47-62.
Clarke, D. 1976. Mesolithic Europe: The economic basis. In *Problems in economic and social archaeology*, ed. G. Sieveking, I. H. Longworth, and K. E. Wilson. London: Duckworth.
Clutton-Brock, J. 1981. *Domesticated animals from early times.* Austin: University of Texas Press.
Coale, A. J. 1972. *The growth and structure of human populations.* Princeton: Princeton University Press.
Coale, A. J. 1974. The history of the human population. *Scientific American* **231** (3):40-51.
Cohen, M. N. 1977. *The food crisis in prehistory.* New Haven: Yale University Press.
Daniel, G. 1967. *The origins and growth of archaeology.* London: Penguin.
De Laet, S. J. 1976. *Acculturation and continuity in Atlantic Europe.* Dissertationes Archaeologicae Gandenses 16. Brugge: De Tempel.
Delibrias, G., J. Evin, and Y. Thommeret. 1982. Sommaire des datations C-14 concernant la préhistoire en France II—Dates parues de 1974 à 1982 Chapitre VI: Néolithique de environ 7000 BP à environ 4000 BP. *Bulletin de la Société Préhistorique Française* **79** (6):175-192.
Dennell, R. 1978. *Early farming in South Bulgaria from the VI to the III millennia B.C.* British Archaeological Reports. International Series 545. Oxford.
Dennell, R. W. 1979. Prehistoric diet and nutrition: Some food for thought. *World Archaeology* **11**:121-135.
Dolukhanov, P. M. 1979. *Ecology and economy in Neolithic Eastern Europe.* London: Duckworth.
Durham, W. H. 1984. *Coevolution: Genes and culture in human populations.* Stanford: Stanford University Press.
Eckert, J., M. Ihmig, A. Jürgens, R. Kuper, H. Löhr, J. Lüning, and I. Schroter. 1971. Untersuchungen zur neolithischen Besiedlung der Aldenhovener Platte. *Bonner Jahrbücher* **171**:558-664.
Eckert, J., M. Ihmig, R. Kuper, H. Löhr, and J. Lüning. 1972. Untersuchungen zur neolithischen Besiedlung der Aldenhovener Platte II. *Bonner Jahrbücher* **172**:344-394.

Edmonson, M. S. 1961. Neolithic diffusion rates. *Current Anthropology* **2**:71-102.
Fairbridge, R. W., and C. Hillaire-Marcel. 1977. An 8,000 year palaeoclimatic record of the "Double Hale" 45 year solar cycle. *Nature* **268**:413-414.
Farruggia, J. P., R. Kuper, J. Lüning, and P. Stehli. 1973. *Der Bandkeramische Siedlungsplatz Langweiler 2.* Rheinische Ausgrabungen 13. Bonn: Rheinland Verlag GMBH.
Fisher, R. A. 1937. The wave of advance of advantageous genes. *Annals of Eugenics, London* **7**:355-369.
Flannery, K. V. 1973. The origins of agriculture. *Annual Review of Anthropology* **2**:271-310.
Geddes, D. 1981. Les débuts de l'élevage dans la vallée de l'Aude. *Bulletin de la Société Préhistorique Française* **78** (10-12):370-378.
Geist, V. 1971. *Mountain sheep.* Chicago: University of Chicago Press.
Gimbutas, M. 1970. Proto-Indo-European culture: The Kurgan Culture during the fifth, fourth, and third millennia B.C. In *Indo-European and Indo-Europeans*, ed. G. Cardona, H. M. Hoenigswald, and A. Senn. Philadelphia: University of Pennsylvania Press.
Gimbutas, M. 1973. The beginning of the bronze age in Europe and the Indo-Europeans: 3500-2500 B.C. *The Journal of Indo-European Studies* **3**:163-214.
Gimbutas, M. (ed.). 1976. *Neolithic Macedonia, as reflected by excavations at Anza, southeast Yugoslavia.* Monumenta Archaeologica. Los Angeles: University of California.
Goodenough, W. H. 1970. The evolution of pastoralism and Indo-European origins. In *Indo-European and Indo-Europeans*, ed. G. Cardona, H. M. Hoenigswald, and A. Senn. Philadelphia: University of Pennsylvania Press.
Gorman, C. 1971. The Hoabinhian and after: Subsistence patterns in Southeast Asia during the late Pleistocene and early Recent period. *World Archaeology* **2**:300-320.
Gramsch, B. (ed.). 1980. *Mesolithikum in Europa.* Veröffentlichungen des Museums für Ur- und Frühgeschichte 14-15. Potsdam.
Grygiel, R., and P. I. Bogucki. 1981. Early neolithic sites at Brześć Krejawski, Poland: Preliminary report on the 1976-1979 excavations. *Journal of Field Archaeology* **8**:9-27.
Guilaine, J. (ed.). 1976. *La Préhistoire française. Les civilisations néolithiques et protohistoriques de la France.* Paris: CNRS.
Guilaine, J. (ed.). 1979a. *L'abri Jean Cros.* Toulouse: Centre d'Anthropologie des Sociétés Rurales.
Guilaine, J. 1979b. The earliest neolithic in the west Mediterranean: A new appraisal. *Antiquity* **53**:22-30.
Hallam, B. R., S. E. Warren, and C. Renfrew. 1976. Obsidian in the western Mediterranean: Characterization by neutron activation analysis and

optical emission spectroscopy. *Proceedings of the Prehistoric Society* **42**:85-110.
Halstead, P. 1981. Counting sheep in neolithic and bronze age Greece. In *Patterns of the past*, ed. I. Hodder, G. Isaac, and N. Hammond. Cambridge: Cambridge University Press.
Hammond, F. 1981. The colonization of Europe: The analysis of settlement processes. In *Patterns of the past*, ed. I. Hodder, G. Isaac, and N. Hammond. Cambridge: Cambridge University Press.
Hansen, J., and J. M. Renfrew. 1978. Palaeolithic-neolithic seed remains at Franchthi Cave, Greece. *Nature* **271**:349-352.
Harlan, J. R. 1971. Agricultural origins: Centers and non-centers. *Science* **174**:468-474.
Harlan, J. R. 1975. *Crops and man*. Madison: American Society of Agronomy.
Hassan, F. 1981. *Demographic archaeology*. New York: Academic Press.
Helbaek, H. 1969. Plant collecting, dry-farming, and irrigation agriculture in prehistoric Deh Luran. In *Prehistory and human ecology of the Deh Luran Plain*, ed. F. Hole, K. Flannery, and J. Neely. Memoirs of the Museum of Anthropology, no. 1. Ann Arbor: University of Michigan.
Hewlett, B., J.M.H. Van de Koppel, and L. L. Cavalli-Sforza. 1982. Exploration ranges of Aka pygmies of the Central African Republic. *Man* **17**:418-430.
Hiernaux, J. 1974. *The people of Africa*. London: Weidenfeld and Nicolson.
Higgs, E. S. (ed.). 1975. *Palaeoeconomy*. Cambridge: Cambridge University Press.
Higgs, E. S., and M. R. Jarman. 1969. The origins of agriculture: A reconsideration. *Antiquity* **43**:31-41.
Hole, F., K. V. Flannery, and J. A. Neely (eds.). 1969. *Prehistory and human ecology of the Deh Luran Plain*. Memoirs of the Museum of Anthropology, no. 1. Ann Arbor: University of Michigan.
Howell, N. 1979. *Demography of the Dobe !Kung*. New York: Academic Press.
Jacobsen, T. W. 1973. Excavations in the Franchthi Cave, 1969-1971, Part I. *Hesperia* **42**:45-88.
Jacobsen, T. W. 1976. 17,000 years of Greek prehistory. *Scientific American* **234** (6):76-87.
Jarman, M. R., and H. N. Jarman. 1968. The fauna and economy of Early Neolithic Knossos. *British School at Athens* **63**:241-261.
Kendall, D. G. 1948. A form of wave propagation associated with the equation of heat conduction. *Proceedings of the Cambridge Philosophical Society* **44**:591-594.
Kendall, D. G. 1965. Mathematical models of the spread of infection. In *Mathematics and computer science in biology and medicine*. London: Medical Research Council.
Kidd, D. 1980. Barbarian Europe in the first millennium. In *The Cambridge encyclopedia of archaeology*, ed. A. Sherratt. New York: Cambridge University Press.

Kimura, M. 1968. Evolutionary rate at the molecular level. *Nature* **217**:624-626.
Klein, J., J. C. Lerman, P. E. Damon, and T. Linick. 1980. Radiocarbon concentration in the atmosphere: 8,000 year record of variations in tree rings. *Radiocarbon* **22**:950-961.
Klein, J., J. C. Lerman, P. E. Damon, and E. K. Ralph. 1982. Calibration of radiocarbon dates: Tables based on the consensus data of the Workshop on Calibrating the Radiocarbon Time Scale. *Radiocarbon* **24**:103-150.
Klejn, L. S. 1976. Das Neolithikum Europas als ein Ganzes. *Jahreschrift für Mittledeutsche Vorgeschichte* **60**:9-22.
Kozlowski, S. K. (ed.). 1973. *The mesolithic in Europe*. Warsaw: Warsaw University Press.
Kruk, J., 1980. *Gospodavka w Polsce Poludniono-Wschodniej*. Warsaw: Polska Academia Nauk.
Kuper, R., H. Löhr, J. Lüning, and P. Stehli. 1974. Untersuchungen zur neolithischen Besiedlung der Aldenhovener Platte IV. *Bonner Jahrbücher* **174**:424-508.
Kuper, R., and J. Lüning. 1975. Untersuchungen zur neolithischen Besiedlung der Aldenhovener Platte. In *Ausgrabungen in Deutschland*. Mainz: Römisch-Germanisches Museum.
Lee, R. B. 1972. Population growth and the beginnings of sedentary life among the !Kung bushmen. In *Population growth: Anthropological implications*, ed. B. Spooner. Cambridge: MIT Press.
Lee, R. B., and I. DeVore (eds.). 1968. *Man the hunter*. Chicago: Aldine.
Levine, P., and R. E. Stetson. 1939. An unusual case of intragroup agglutination. *Journal of the American Medical Association* **113**:127-129.
McIntosh, S. K., and R. J. McIntosh. 1982. West African prehistory. *American Scientist* **69**:603-613.
McKeown, T. 1976. *The modern rise of population*. New York: Academic Press.
Mellaart, J. 1975. *The neolithic of the Near East*. New York: Scribners.
Mellars, P. (ed.). 1978. *The early postglacial settlement of northern Europe*. London: Duckworth.
Menozzi, P., A. Piazza, and L. L. Cavalli-Sforza. 1978. Synthetic maps of human gene frequencies in Europeans. *Science* **201**:786-792.
Menozzi, P., A. Piazza, and L. L. Cavalli-Sforza. 1984. Methodology and results of analysis of correlations between gene frequencies and climate. Manuscript.
Milisauskas, S. 1978. *European prehistory*. New York: Academic Press.
Modderman, P.J.R. (ed.). 1970. *Linearbandkeramik aus Elsloo and Stein*. Leiden: Analecta Praehistorica Leidensia III.
Moore, A.M.T. 1975. The excavation of Tell Abu Hureyra in Syria: A preliminary report. *Proceedings of the Prehistoric Society* **41**:50-77.

Moore, A.M.T. 1979. A pre-Neolithic farmer's village on the Euphrates. *Scientific American* **241** (2):62-70.
Moore, A.M.T. 1982. Agricultural origins in the Near East: A model for the 1980s. *World Archaeology* **14**:224-236.
Mourant, A. E. 1954. *The distribution of the human blood groups.* Oxford: Blackwell.
Mourant, A. E., A. C. Kopeć, and K. Domaniewska-Sobczak. 1976. *The distribution of the human blood groups and other polymorphisms.* 2nd ed. Oxford: Oxford University Press.
Murray, J. 1970. *The first European agriculture.* Edinburgh: Edinburgh University Press.
Neustupny, E. 1968. Absolute chronology of the Neolithic and Aeneolithic periods in central and south-east Europe. *Slovenská Archeológia* **16**:19-56.
Newell, R. R. 1970. The flint industry of the Dutch Linearbandkeramik. In *Linearbandkeramik aus Elsloo and Stein*, ed. P.J.R. Modderman. Leiden: Analecta Praehistorica Leidensia III.
Olsson, I. U. (ed.). 1970. *Radiocarbon variations and absolute chronology.* New York: Wiley.
Phillips, P. 1975. *Early farmers of west Mediterranean Europe.* London: Hutchinson.
Piazza, A., P. Menozzi, and L. L. Cavalli-Sforza. 1980. The HLA-A,B gene frequencies in the world: Migration or selection? *Human Immunology* **4**:297-304.
Piazza, A., P. Menozzi, and L. L. Cavalli-Sforza. 1981a. The making and testing of geographic gene frequency maps. *Biometrics* **37**:635-659.
Piazza, A., P. Menozzi, and L. L. Cavalli-Sforza. 1981b. Synthetic gene frequency maps of man and selective effects of climate. *Proceedings of the National Academy of Science* **78** (4):2638-2642.
Randsborg, K. 1975. Social dimensions of early neolithic Denmark. *Proceedings of the Prehistoric Society* **41**:105-118.
Redman, C. L. 1978. *The rise of civilization.* San Francisco: Freeman.
Reed, C. A. (ed.). 1977. *Origins of agriculture.* The Hague: Mouton.
Rendine, S., A. Piazza, and L. L. Cavalli-Sforza. 1984. Simulation of the spread of early farming. In preparation.
Renfrew, C. 1973. *Before civilization: The radiocarbon revolution and prehistoric Europe.* London: Jonathan Cape.
Renfrew, C., J. E. Dixon, and J. R. Cann. 1968. Further analysis of Near Eastern obsidians. *Proceedings of the Prehistoric Society* **34**:319-331.
Renfrew, J. M. 1973. *Palaeoethnobotany.* New York: Columbia University Press.
Roughgarden, J. 1979. *Theory of population genetics and evolutionary ecology: An introduction.* New York: Macmillan.
Ryder, L. P., E. Andersen, and A. Sverjgaard. 1978. An HLA map of Europe. *Human Heredity* **28**:171-200.

Sabloff, J. A. (ed.). 1981. *Simulations in archaeology*. Albuquerque: University of New Mexico Press.
Sgaramella-Zonta, L., and L. L. Cavalli-Sforza. 1973. A method for the detection of a demic cline. In *Genetic structure of population*, ed. N. E. Morton. Population Genetics Monograph, vol. 3. Honolulu: University Press of Hawaii.
Sillen, A., and M. Kavanagh. 1982. Strontium and paleodietary research: A review. *The Yearbook of Physical Anthropology* **25**:67-90.
Simoons, F. J. 1978. The geographic hypothesis and lactose malabsorption. *American Journal of Digestive Disease* **23**:963-980.
Simoons, F. J. 1981. Celiac disease as a geographic problem. In *Food, nutrition and evolution*, ed. D. N. Walcher and N. Kretchmer. New York: Masson Publishing.
Skellam, J. 1951. Random dispersal in theoretical populations. *Biometrika* **38**:196-218.
Skellam, J. 1973. The formulation and interpretation of mathematical models of diffusionary processes in population biology. In *The mathematical theory and dynamics of biological populations*, ed. M. Bartlett and R. Hiorns. London: Academic Press.
Smith, P.E.L. 1975. Ganj Dareh Tepe. *Iran* **13**:178-180.
Sokal, R. R., and P. Menozzi. 1982. Spatial autocorrelations of HLA frequencies in Europe support demic diffusion of early farmers. *American Naturalist* **119**:1-17.
Sokal, R. R., and F. J. Rohlf. 1981. *Biometry*. 2nd ed. San Francisco: Freeman.
Soudsky, B., and I. Pavlu. 1972. The linear pottery culture settlement patterns in central Europe. In *Man, settlement and urbanism*, ed. P. Ucko, R. Tringham, and G. Dimbleby. London: Duckworth.
Spooner, B. (ed.). 1972. *Population growth: Anthropological implications*. Cambridge: MIT Press.
Startin, W. 1978. Linear pottery culture houses: Reconstruction and manpower. *Proceedings of the Prehistoric Society* **44**:143-160.
Stauder, J. 1971. *The Majangir: Ecology and society of a southwest Ethiopian people*. Cambridge: Cambridge University Press.
Sutter, J., and Tran Ngoc Toan. 1957. The problem of the structure of isolates and of their evolution among human populations. *Cold Spring Harbor Symposia on Quantitative Biology* **22**:379-383.
Tauber, H. 1981. ^{13}C evidence for dietary habits of prehistoric man in Denmark. *Nature* **292**:332-333.
Tringham, R. 1971. *Hunters, fishers and farmers of eastern Europe 6000-3000 B.C.* London: Hutchinson.
Ucko, P. J., and G. W. Dimbleby (eds.). 1969. *The domestication and exploitation of plants and animals*. London: Duckworth.
Uerpmann, H. P. 1979. *Probleme der Neolithisierung des Mittelmeerraumes*.

Beihefte zum Tubinger Atlas des Vorderen Orients, ser. B, 28. Wiesbaden: Reichert Verlag.
Ulbrich, J. 1930. *Die Bisamratte*. Dresden: Heinrich.
Van de Velde, P. 1979. *On Bandkeramik social organization*. Analecta Praehistorica Leidensia 12. The Hague: Leiden University Press.
Van Zeist, W. 1976. On macroscopic traces of food plants in southwestern Asia. In *The early history of agriculture*. Philosophical Transactions of the Royal Society of London B, vol. 275, no. 936.
Van Zeist, W., and J.A.H. Bakker-Heeres. 1979. Some economic and ecological aspects of the plant husbandry of Tell Aswad. *Paleorient* **5**:161-169.
Van Zeist, W., and S. Bottema. 1981. Palynological evidence for the climatic history of the Near East, 50,000-6,000 B.P. In *Préhistoire du Levant*. Colloques Internationaux du CNRS, no. 598. Paris.
Vavilov, N. I. 1926. *Studies on the origin of cultivated plants*. Leningrad.
Waterbolk, H. T. 1968. Food production in prehistoric Europe. *Science* **162**:1093-1102.
Weiss, K. 1973. *Demographic models in anthropology*. Memoirs of the Society for American Archaeology, no. 27. Washington, D.C.
Wendorf, F., and R. Schild (eds.). 1980a. *Prehistory of the eastern Sahara*. New York: Academic Press.
Wendorf, F., and R. Schild (eds.). 1980b. *Loaves and fishes: The prehistory of Wadi Kubbaniya*. Dallas: Department of Anthropology, Southern Methodist University.
Wishart, D. J., A. Warren, and R. H. Stoddard. 1969. An attempted definition of a frontier using a wave analogy. *The Rocky Mountain Social Science Journal* **6**:73-81.
Wright, H. E. 1977. Environmental change and the origin of agriculture in the Old and New Worlds. In *Origins of agriculture*, ed. C. A. Reed. The Hague: Mouton.
Zohary, D. 1969. The progenitors of wheat and barley in relation to domestication and agricultural dispersal in the Old World. In *The domestication and exploitation of plants and animals*, ed. P. Ucko and G. Dimbleby. London: Duckworth.
Zohary, D. 1970. Centers and diversity and centers of origin. In *Genetic resources in plants: Their exploration and conservation*, ed. O. Frankel and E. Bennett. Philadelphia: F. A. Davis.
Zohary, D., and M. Hopf. 1973. Domestication of pulses in the Old World. *Science* **182**:887-894.
Zvelebil, M. 1980. The rise of the nomades in Central Asia. In *The Cambridge encyclopedia of archaeology*, ed. A. Sherratt. New York: Cambridge University Press.

INDEX

ABO gene system, 7, 94-98, 104, 107, 119, 136
aborigines, 4, 9, 153n5.1
Abri Jean Cros, 40 (Fig. 3.3), 41 (Tab. 3.1)
Abu Hureyra, 20 (Tab. 2.1), 26, 30, 149
Acconia, 151n3.19
acculturation, 62, 117, 118, 126. *See also* population, interactions
Aegean Sea, 41
Aegilops squarrosa, 20
African rice, 15
age structure, 23. *See also* population
agriculture, 4, 9-10, 30, 39, 139; Andes, 15; China, 14; Europe, 39-45; Mesoamerica, 14; modern, 10; Near East, 14, 30-32; Sub-Saharan Africa, 15; yields, 10, 27. *See also* agriculture, origins of; domestication
agriculture, origins of, 6, 9, 10-12, 13, 16, 32, 151n3.17; Andes, 15; China, 14; Mesoamerica, 15; Near East, 14, 20 (Tab. 2.1), 30-31; Sub-Saharan Africa, 15; theories, 11, 27-30, 149n2.14. *See also* agriculture; domestication
Alaska, 153n5.1
Aldenhoven Platte, 40 (Fig. 3.3), 43, 44 (Fig. 3.5), 63, 73, 75 (Fig. 5.6), 154n5.11, 157n7.3
Algonkian, 153n5.1
Ali Kosh, 20 (Tab. 2.1), 30-31, 57
Alps, 36, 55, 58, 134
Andean Highlands, 15
anemia, 86
anthropomorphic figurines, 42
Anza, 41 (Tab. 3.1)
AP gene system, 99
Apulia, 45
Arabian Peninsula, 138
Argissa, 41 (Tab. 3.1)
artificial selection, 22, 32
Ashkenazi Jews, 98
Asiab, 31
Aswad, Tell, 4, 20 (Tab. 2.1), 30, 149n2.14
Azmak, 40 (Fig. 3.3)

Balkans, 42, 43, 50, 57, 58, 78, 149n1.3, 150n3.9
Balma Abeurador, 150n3.7, 151n3.17
Baltic Sea, 34, 39, 151n3.18
BANDK 2 model, 113, 114
Bandkeramik culture, 42, 43, 44 (Fig. 3.5), 45, 58, 62, 73-77, 113-116, 134, 154n5.11, 157n7.3
Bantu, 117, 139
barley, 6, 14, 19, 20, 26, 30, 39, 45, 133, 151n3.17
Basi, 150n3.13
Basques, 86, 92, 156n6.4
Battle Axe culture, 108
beans, 15
Beidha, 20 (Tab. 2.1), 30-31
Binford, 29, 149n2.14
birth interval, 64
birth rates, 63, 66. *See also* population
Black Sea, 42, 108, 158n7.9
boats, 44, 134
Braidwood, R. J., 11, 12, 14, 28
bread wheat, 19
Breść Kujawski, 41 (Tab. 3.1)
bristlecone pine, 56
British Isles, 45, 58, 60
broad spectrum economies, 27
Bulgaria, 40
bushmen, 4, 9, 81, 153n5.1
Bylany, 40 (Fig. 3.3), 43, 113

Calabria, xiii, 45, 135, 150n3.12, 151n3.19
California, 153n5.1
Cameroon, 139

Canada, 153n5.1
Cap Ragnon, 150n3.13
carbon-14 dating. *See* radiocarbon dating
carrying capacity, 63, 72
Catal Hüyük, 31
cattle, 6, 24, 40-41, 42-43, 137
Caucasus, 138
Çayönü, 20 (Tab. 2.1), 30-31, 57
Central African Republic, 78
Châteauneuf-les-Martiques, 38
Childe, V. G., 3, 11, 12, 16, 28, 34, 35
chili pepper, 15
China, 13, 14
Chipewyan, 153
Clark, D., 39
Clark, J. D., 16
Clark, J.G.D., 50
climatic change, 28-29, 36ff.
cline, 85, 94. *See also* demic cline
Cohen, M. N., 16
computer simulation. *See* simulation studies
contour maps. *See* principal components analysis
correlation coefficient, 54, 57, 106
Corsica, 150n3.13
Cova con Ballester, 151n3.13
Cova Fosca, 150n3.13
Covet de l'Or, 40 (Fig. 3.3)
Criş culture, 42
cross-pollination, 19
Cueva Matutano, 151n3.13
Cueva del Parco, 150n3.13
cultural diffusion, 6, 61-62, 82, 134
Curacchiaghiu, 150n3.13
Czechoslovakia, 43, 113

Danube River, 39, 42, 48, 150
Danubian culture. *See* Bandkeramik culture
de Candolle, A., 10, 11, 12
demic cline, 128-130
demographic transition, 66
dendrochronology, 56
Denmark, 34, 39, 45, 58, 150n3.2, 151n3.21
density equilibrium model, 29
diffusion, xiv, 6-7, 61-62, 67-68, 134.

See also cultural diffusion; demic cline
diffusionism, xiv, 6
Divostin, 40 (Fig. 3.3)
DNA, 87-89
dog, 23, 25, 31
domestication, 6, 12, 16; plant, 16-19; animal, 21-23

Edmonson, M. S., 52
Egypt, 24
einkorn, 19, 20, 21, 26-27, 151n3.17. *See also* wheat
Ellerbek culture, 39
emmer, 19, 20, 43, 52, 133. *See also* wheat
England, 34, 36, 38, 134
Ertebølle culture, 39, 141, 150n3.2, 151n3.18
Ethiopia, 80, 155n5.15
ethnographic studies, 63, 78
Euphrates River, 26
exploration range, 78
exponential growth, 71-72, 157n7.3

finger millet, 15
fish, 24, 38-39
Fisher, R. A., 68, 69, 110, 135, 154n5.7
fitness. *See* genetic fitness
Flannery, K. V., 27, 149n2.14
flax, 31
Fontbregoua, 151nn3.7, 3.17
food production. *See* agriculture; agriculture, origins of
forest neolithic cultures, 150n3.2
foxtail millet, 13, 14
France, 36, 43-45, 46, 86, 150nn3.7, 3.13, 151nn3.17, 3.18
Franchthi Cave, 39, 52, 149n1.2, 151n3.17
frontier zone, 61, 73, 74 (Fig. 5.5). *See also* wave of advance model
Fy gene system, 99

Ganj Dareh, 31
Gaussian surface, 76
gene frequencies. *See* genes
genes, xiv, 7, 16, 68, 83, 85, 126, 132, 136-137; cline, 85, 94; DNA, 87;

INDEX

frequencies, 7, 82-83, 85, 86, 139; genetic distance, 100; maps of genes, 88 (Fig. 6.1), 95-100 (Figs. 6.4–6.8); mutation, 87, 89; natural selection, 90, 101; neutrality vs. selection, 91; random genetic drift, 87, 120; simulation of gene pools, 119-120; synthetic gene maps, 99, 105-107 (Figs. 6.10–6.12). *See separate entries for individual gene systems*
genetic drift, 89, 101, 120, 122 (Fig. 7.5), 126, 158nn7.7, 7.10
genetic fitness, 90
Germany, 43, 45, 63, 73
giant sequoia, 56
glaciers, 36
gluten intolerance, 137, 158n8.1
goats, 6, 21-23, 31, 41, 42, 151n3.17
grape, 13
Greece, 10, 39, 43, 47, 50, 52, 57, 58, 60, 78, 87, 133, 149n1.2, 151nn3.17, 3.18
Greek colonization, 61
grinding equipment, 25-28
Grotta della Madonna, 38 (Fig. 3.2), 39, 150n3.12, 151n3.17
Grotta dell'Uzzo, 151n3.17
Grotta di Sant'Angelo, 150n3.12
growth rates. *See* population
Guran, Tepe, 31
Gyálarét, 41 (Tab. 3.1)

Haçilar, 20 (Tab. 2.1), 30
Haida, 153n5.1
Hardy-Weinberg laws, 156n6.4
Harlan, J. R., 13, 27, 29
Hemudu, 17
HLA gene system, 94, 98, 99, 100 (Fig. 6.8), 106-107, 137, 158n8.1
Holland, 36, 43, 153n4.5, 157n7.3
Holocene, 16, 36
Hp gene system, 99
Hungary, 42, 43, 151n3.17
hunter-gatherers, 4, 5 (Fig. 1.1), 9, 24, 59, 62, 63, 119, 120, 153n5.1
hybridization, 19

Iberian Peninsula, 58
Ile Riou, 150n3.13

indigenism, xiv
Indo-European languages, 86, 108, 157n6.13
Indus Valley, 138
Iran, 21, 31, 133, 138
Iraq, 23
Iraqi Kurdistan, 24-25
isochron maps, 50, 58, 152nn4.2, 4.4; Neolithic map, 53, 58; Mesolithic map, 54, 59
Issongos, 78
Italy, 39, 43, 46, 87, 135, 150n3.12, 151n3.17

Jarmo, 57
Jericho, 20 (Tab. 2.1), 30-31, 53 (Fig. 4.2), 57
Jordan Valley, 21
Judea, 12
Jukes, T. A., 92

Karanovo culture, 41
Kebaran period, 26
Kendall, D. G., 68
Kiel-Ellerbek, 38 (Fig. 3.2)
Kimura, M., 92
King, J. L., 92
Knossos, 40 (Fig. 3.3), 41 (Tab. 3.1), 151n3.18
Körös culture, 42
Kremikovci culture, 41

lactose absorption, 137, 158n8.2
Languedoc, 150n3.13
Lautereck, 38 (Fig. 3.2)
Lebanon, 12
Lee, R. B., 64
Le gene system, 99
lentil, 14, 20 (Tab. 2.1), 30, 150n3.7
Lepenski Vir, 38 (Fig. 3.2), 39, 48, 150n3.6
Levant, 21, 22, 24, 26, 31
Levine, P., 86
Libby, W. F., 50
life tables, 64. *See also* population
Linear pottery culture. *See* Bandkeramik culture
logistic growth, 68, 71-72, 116, 154n5.10, 155n5.12

Lotka-Volterra equations, 116-118, 157n7.4
Lubbock, J., 34

Macedonia, 41
MacNeish, R. S., 15
macro-simulation. *See* simulation studies
maize, 10, 12, 13, 15, 19
Majangir, 80, 156n5.17
map construction. *See* principal components analysis
mating distances, 76, 78, 79 (Fig. 5.8)
measurement of rate of spread, 51-58, 134
Mediterranean, 36, 39, 43, 58, 61, 134, 150n3.9
Menozzi, P., 104
Merzbach, 43, 44 (Fig. 3.5), 73
Mesoamerica, 12, 13, 14
mesolithic, 34-39, 46, 59, 62; C-14 dates of sites, 146; environmental adaptation, 36; isochron map of "latest" occupation, 60 (Fig. 4.6); settlement patterns, 39, 150n3.6; stone tools, 36; subsistence, 37-38, 150n3.7
Mesopotamia, 10
microliths, 34, 36, 37 (Fig. 3.1)
micro-simulation. *See* simulation studies
migratory activity, 68, 76, 111, 115-116, 123, 135, 155n5.15, 158n7.8
mixed model, 83
MNS gene system, 99
model building, 67
Moita do Sabastião, 38 (Fig. 3.2), 39
Montagnais, 153n5.1
mortality rates, 63-66. *See also* population
Mourant, A. E., 86
Mureybat, 20 (Tab. 2.1), 26, 30, 149
muskrat, spread of, 69-70
mutation, 87-89. *See* genes
mutualism, 62, 117, 139. *See also* population

Natufian culture, 26
natural habitat zone, 11
natural selection, 84, 86, 87, 90-92, 101, 132, 137

Nea Nikomedia, 40 (Fig. 3.3)
Negev, 26
neolithic, 3, 30-33, 34-35, 50; building activity, 35, 48; C-14 dates of sites, 142; early cultures in Europe, 39-45; history of the term, 34, 50; map of selected sites in Europe, 40; neolithic rise of population, 63-67; rate of spread in Europe, 53, 57, 58; stone tools, 46; subsistence, 30-33, 35, 40-43, 47; transition, 3, 6, 45-49, 60-62, 63, 138. *See also* pottery; settlement patterns
neolithic revolution, 3, 34
neutrality, 91
Nigeria, 139
Nile Valley, 24, 138, 149n2.1
North Africa, 138, 153n5.1
North Sea, 41
Nubia, 24

obsidian exchange, 32, 45
Oder River, 45
olive, 13

paleobotanical studies, 20, 52. *See also* pollen analysis
paleolithic, 34, 36
Pavlu, I., 113
pearl millet, 15
peas, 14, 20 (Tab. 2.1), 30, 150n3.7
Piana di Curinga, 40 (Fig. 3.3), 45
Piazza, A., 104
pigs, 6, 14, 21-23, 31, 40
pistachio, 20 (Tab. 2.1)
Pleistocene, 16, 35, 36, 149n2.14
PMG gene system, 99
Poland, 45, 108, 157n7.3
pollen analysis, 28, 37
polypoidy, 19
population, 3, 4, 7, 116, 138; density, 63, 115; doubling times, 71; growth, 32, 63, 66, 71, 72, 116, 126, 136; interactions, 8, 59, 62, 116-119, 134; neolithic rise, 63-66; pressure, 16, 29; rates of growth, 63, 71, 73-76, 80-81, 114, 116, 135; simulation, 119-126; world today, 4, 9. *See also* birth rates; exponential growth; logistic growth; Lotka-Volterra equa-

INDEX

tions; migratory activity; mortality rates
Portugal, 39
pottery, 34, 39-45; Bandkeramik, 42, 43; cardial, 42 (Fig. 3.4), 150n3.13, 151n3.18; Criş, 42 (Fig. 3.4); Ertebølle, 150n3.2, 151n3.18; Grimston–Lyles Hill, 45; impresso, 44; Sesklo, 41; TRB, 45
principal axis, 152
principal components analysis, 102-105, 132; contour map of first component, 105; contour map of second component, 106 (Fig. 6.11), 107; contour map of third component, 107, 108; contour map of macro-simulation, 130, 131 (Fig. 7.10)
propinquity theory, 11
Provence, 150n3.13
pygmies, 4, 9, 78, 117, 153n5.1
Pyrenees, 86

Radiocarbon, 55, 141
radiocarbon dating, xiii, 50, 55, 58, 152n4.2; calibration, 55, 152n4.2; list of neolithic dates, 142; list of mesolithic dates, 146
Ramad, Tell, 30
rate of spread, 51-58, 134
regression analysis, 53 (Fig. 4.2), 54-57, 152nn4.2, 4.3
regression line, 54, 55-57
Renfrew, C., 149n1.3
Rh gene system, 7, 86-89, 92-93, 94, 104, 121, 136, 156nn6.3, 6.4
rice, 10, 14
Roman Empire, 108
Rome, 10
Rumania, 42, 58

Sahara, 15, 138
Sauveterrian, 36
Scandinavia, 36
sea-level changes, 36
sedentism, 25-28, 32, 48
selective advantage, 90
self-pollination, 19
Sesklo culture, 41
Setaria, 15

settlement patterns, 39, 43, 47-48, 77, 113-114
Sgaramella-Zonta, L., 100, 128
Shanidar Cave, 25
sheep, 6, 21-23, 31, 40-43, 151n3.17
shellfish, 38-39
Shensi, 14
simulation studies, 8, 68, 110-113; macro-simulation, 113, 119-126, 137; micro-simulation, 112, 116, 135
site catchment studies, 151n3.19
skeletal remains, 64, 65 (Fig. 5.1), 138, 151n3.21
Skellam, J. G., 68, 69, 77, 153n5.7
sorghum, 15
Soudsky, B., 113
Soviet Union, 141, 150nn3.2, 3.9, 151n3.18
Spain, 43, 86, 150n3.13
Spirit Cave, 14, 149n2.1
spread of early farming. *See* neolithic; rate of spread
stand management model, 30
Star Carr, 38
Starčevo culture, 42
stepping stone model, 111
Stetson, R., 86
stochastic processes, 112
stone tools. *See* mesolithic; neolithic
storage pits, 26-28
swidden farming, 43, 113
synthetic gene maps, 7-8, 99-108, 136, 137. *See also* principal components analysis
Syria, 31

Tardenoisian, 36
Taurus-Zagros region, 12, 21
Tavoliere, 45
Tehuacan Valley, 15
teosinte, 15
Thailand, 14
thalassemia, 99
Thessaly, 42
Thomsen, C. J., 34
three age system, 34
time series analysis, 107
TRB culture, 45
Turkey, 27, 31, 32, 87, 134

Ukraine, 58, 108

Vavilov, N. I., 10, 11, 12
Vavilovian centers, 10
vetch, 31

Wadi Kubbaniya, 149n2.1
warfare, 117. *See also* population
wave of advance model, xiv, 6, 61-62, 68-69, 110, 135; formulation, 68-69, 153-154nn5.7, 5.8; genetic implications, 82-84; growth component, 71-76, 154-155nn5.10–5.12; initial test, 80-81, 156n5.17; migratory component, 76-80, 155nn5.15, 5.16; simulation of, 110, 112-116, 119-128; wave front, 68, 77, 153n5.7
Wendorf, F., 24
western Mediterranean, 43, 61, 134, 150n3.9
wheat, 6, 10, 14, 17, 19, 30, 40, 46, 52. *See also* einkorn, emmer
wild progenitors, 20, 52. *See also* barley; wheat
Worsaae, J.J.A., 34

yams, 15
Yang-shao culture, 14
Yugoslavia, 41, 42

Zawi Chemi Shanidar, 23, 25, 31

Library of Congress Cataloging in Publication Data

Ammerman, Albert J., 1942-
   The neolithic transition and the genetics of
populations in Europe.

   Bibliography: p.   Includes index.
   1. Neolithic period—Europe. 2. Human population
genetics—Europe. 3. Man, Prehistoric—Europe—
Population. 4. Agriculture—Origin.
5. Europe—Antiquities.
I. Cavalli-Sforza, Luigi Luca, 1922-   . II. Title.
GN776.2.A1A46 1985   304.5′0936   84-42587
ISBN 0-691-08357-6 (alk. paper)